A CENTURY OF ECOSYSTEM SCIENCE

Planning Long-Term Research in the Gulf of Alaska

Committee to Review the Gulf of Alaska Ecosystem Monitoring Program

Polar Research Board

Board on Environmental Studies and Toxicology

Division on Earth and Life Studies

National Research Council

NATIONAL ACADEMY PRESS
Washington, D.C.

NATIONAL ACADEMY PRESS • 2101 Constitution Avenue, N.W. • Washington, DC 20418

NOTICE: The project that is the subject of this report was approved by the Governing Board of the National Research Council, whose members are drawn from the councils of the National Academy of Sciences, the National Academy of Engineering, and the Institute of Medicine. The members of the committee responsible for the report were chosen for their special competences and with regard for appropriate balance.

This study was supported by Contract/Grant No. CMRC/WASC/NOAA 50ABNF-0-00013 (BAA00360) between the National Academy of Sciences and *Exxon Valdez* Oil Spill Trustee Council. Any opinions, findings, conclusions, or recommendations expressed in this publication are those of the author(s) and do not necessarily reflect the views of the organizations or agencies that provided support for the project.

International Standard Book Number: 0-309-08473-3

Copies of this report are available from:
Polar Research Board, TNA 751
500 5th Street, NW
Washington, DC 20001
202-334-3479

or

National Academy Press
2101 Constitution Avenue, NW
Lockbox 285
Washington, DC 20055
800-624-6242
202-334-3313 (in the Washington metropolitan area)
http://www.nap.edu

Cover: The background is a SeaWiFS satellite image showing the brilliant phytoplankton bloom in the Gulf of Alaska. This photo was provided by the SeaWiFS Project, National Aeronautics and Space Administration/Goddard Space Flight Center, and ORBIMAGE. The second image, the inlay of the two Orcas, *Orcinus orca*, affectionately known as Maverick and his wingman Iceman, was taken in June 1992 by Retired Commander John Bortniak of the National Oceanic and Atmospheric Administration (NOAA) Corps. Photo courtesy of NOAA photo library. Cover design by Van Nguyen of the National Academy Press.

Printed in the United States of America

Copyright 2002 by the National Academy of Sciences. All rights reserved.

THE NATIONAL ACADEMIES

National Academy of Sciences
National Academy of Engineering
Institute of Medicine
National Research Council

The **National Academy of Sciences** is a private, nonprofit, self-perpetuating society of distinguished scholars engaged in scientific and engineering research, dedicated to the furtherance of science and technology and to their use for the general welfare. Upon the authority of the charter granted to it by the Congress in 1863, the Academy has a mandate that requires it to advise the federal government on scientific and technical matters. Dr. Bruce M. Alberts is president of the National Academy of Sciences.

The **National Academy of Engineering** was established in 1964, under the charter of the National Academy of Sciences, as a parallel organization of outstanding engineers. It is autonomous in its administration and in the selection of its members, sharing with the National Academy of Sciences the responsibility for advising the federal government. The National Academy of Engineering also sponsors engineering programs aimed at meeting national needs, encourages education and research, and recognizes the superior achievements of engineers. Dr. Wm. A. Wulf is president of the National Academy of Engineering.

The **Institute of Medicine** was established in 1970 by the National Academy of Sciences to secure the services of eminent members of appropriate professions in the examination of policy matters pertaining to the health of the public. The Institute acts under the responsibility given to the National Academy of Sciences by its congressional charter to be an adviser to the federal government and, upon its own initiative, to identify issues of medical care, research, and education. Dr. Harvey V. Fineberg is president of the Institute of Medicine.

The **National Research Council** was organized by the National Academy of Sciences in 1916 to associate the broad community of science and technology with the Academy's purposes of furthering knowledge and advising the federal government. Functioning in accordance with general policies determined by the Academy, the Council has become the principal operating agency of both the National Academy of Sciences and the National Academy of Engineering in providing services to the government, the public, and the scientific and engineering communities. The Council is administered jointly by both Academies and the Institute of Medicine. Dr. Bruce M. Alberts and Dr. Wm. A. Wulf are chairman and vice chairman, respectively, of the National Research Council.

COMMITTEE TO REVIEW THE GULF OF ALASKA ECOSYSTEM MONITORING PROGRAM

Members

MICHAEL ROMAN, *Chair*, University of Maryland, Cambridge
DON BOWEN, Fisheries and Oceans Canada, Dartmouth, Nova Scotia
ADRIA A. ELSKUS, University of Kentucky, Lexington
JOHN J. GOERING, University of Alaska, Fairbanks
GEORGE HUNT, University of California, Irvine
SETH MACINKO, University of Rhode Island, Narragansett
DONAL MANAHAN, University of Southern California, Los Angeles
BRENDA NORCROSS, University of Alaska, Fairbanks
J. STEVEN PICOU, University of South Alabama, Mobile
THOMAS C. ROYER, Old Dominion University, Norfolk, Virginia
JENNIFER RUESINK, University of Washington, Seattle
KARL TUREKIAN, Yale University, New Haven, Connecticut

Staff

CHRIS ELFRING, Director, Polar Research Board
DAVID POLICANSKY, Associate Director, Board on Environmental Studies and Toxicology
ANN CARLISLE, Administrative Associate

POLAR RESEARCH BOARD

Members

ROBIN BELL, *Chair*, Lamont-Doherty Earth Observatory, Palisades, New York
RICHARD B. ALLEY, Pennsylvania State University, University Park
AKHIL DATTA-GUPTA, Texas A&M University, College Station
HENRY P. HUNTINGTON, Huntington Consulting, Eagle River, Alaska
AMANDA LYNCH, University of Colorado, Boulder
ROBIE MACDONALD, Fisheries and Oceans, Canada, Institute of Ocean Sciences, Sidney, British Columbia
MILES MCPHEE, McPhee Research Company, Naches, Washington
CAROLE L. SEYFRIT, Old Dominion University, Norfolk, Virginia

Ex-Officio Members

MAHLON C. KENNICUTT, Texas A&M University, College Station
ROBERT RUTFORD, University of Texas, Dallas
PATRICK WEBBER, Michigan State University, East Lansing

Staff

CHRIS ELFRING, Director
ANN CARLISLE, Administrative Associate
ROB GREENWAY,[1] Project Assistant

---

[1] Until November 2000.

BOARD ON ENVIRONMENTAL STUDIES AND TOXICOLOGY

Members

GORDON ORIANS (*Chair*), University of Washington, Seattle
JOHN DOULL (*Vice Chair*), University of Kansas Medical Center, Kansas City
DAVID ALLEN, University of Texas, Austin
INGRID C. BURKE, Colorado State University, Fort Collins
THOMAS BURKE, Johns Hopkins University, Baltimore, Maryland
WILLIAM L. CHAMEIDES, Georgia Institute of Technology, Atlanta
CHRISTOPHER B. FIELD, Carnegie Institute of Washington, Stanford, California
DANIEL S. GREENBAUM, Health Effects Institute, Cambridge, Massachusetts
BRUCE D. HAMMOCK, University of California, Davis
ROGENE HENDERSON, Lovelace Respiratory Research Institute, Albuquerque, New Mexico
CAROL HENRY, American Chemistry Council, Arlington, Virginia
ROBERT HUGGETT, Michigan State University, East Lansing
JAMES H. JOHNSON, Howard University, Washington, D.C.
JAMES F. KITCHELL, University of Wisconsin, Madison
DANIEL KREWSKI, University of Ottawa, Ottawa, Ontario
JAMES A. MACMAHON, Utah State University, Logan
WILLEM F. PASSCHIER, Health Council of the Netherlands, The Hague
ANN POWERS, Pace University School of Law, White Plains, New York
LOUISE M. RYAN, Harvard University, Boston, Massachusetts
KIRK SMITH, University of California, Berkeley
LISA SPEER, Natural Resources Defense Council, New York, New York

Senior Staff

JAMES J. REISA, Director
DAVID J. POLICANSKY, Associate Director and Senior Program Director for Applied Ecology
RAYMOND A. WASSEL, Senior Program Director for Environmental Sciences and Engineering
KULBIR BAKSHI, Program Director for the Committee on Toxicology
ROBERTA M. WEDGE, Program Director for Risk Analysis
K. JOHN HOLMES, Senior Staff Officer
SUSAN MARTEL, Senior Staff Officer
SUZANNE VAN DRUNICK, Senior Staff Officer
RUTH E. CROSSGROVE, Managing Editor

Preface

This report is in response to a request from the Exxon Valdez Oil Spill Trustee Council to review the Gulf of Alaska Ecosystem Monitoring and Research Program (GEM). To ensure that the GEM program is based on a science plan that is robust, far-reaching, and scientifically sound, the Trustee Council asked the National Academies to serve as an independent advisor. The Academies appointed a special committee and charged it to review the scope and content of the program as it evolved. To meet this charge our committee reviewed Trustee Council planning documents and met with their representatives and with individuals representing various communities and user groups of the Gulf of Alaska region.

Trustee Council funds for long-term research in the Gulf of Alaska provide a rare opportunity for citizens, resource managers, and scientists to understand an ecosystem and obtain data essential to its long-term management. Virtually all ecosystems on Earth are influenced by natural changes and human activities. Sustained observations are necessary to separate the influences of these factors and to document natural fluctuations of ecosystem processes. We face this challenge in managing the living resources of all ecosystems. Thus the financial commitment to GEM, if coupled with careful planning and sound science, can serve as a model for ecosystem science and management. This is an exciting prospect.

This report is not an endorsement of a specific science plan for the long-term study of the Gulf of Alaska. While planning is well under way, the details of such a plan will arise after careful analysis, synthesis, and scientific deliberation. We focus this review on the planning process and scientific infrastructure necessary for a successful long-term environmental research program in the Gulf. We make recommendations on how the GEM planning process can be improved, based on the experience of the committee and lessons learned from other environmental research programs. Our report is divided into sections relating to planning long-term ecosystem science; the importance of a conceptual foundation; determining scope and geographic focus; organization structure; community involvement and traditional knowledge; data management; and synthesis, modeling, and evaluation. We recommend a course of action that has proven successful in planning and implementing other large interdisciplinary science programs.

Many people provided information to this committee as we prepared our report. In particular we would like to thank Molly McCammon, Phil Mundy, and Robert Spies of the Trustee Council; Gary Kompkoff from the village of Tatitlek; and Patty Brown-Schwalenberg of the Chugach Regional Resources Commission. On behalf of the entire committee I want to thank Chris Elfring of the Polar Research Board and David Policansky of the Board on Environmental Studies and Toxicology. Their sage council, broad experience with the NRC process, diligence, and professionalism greatly contributed to this report. We thank Ann Carlisle of the Polar Research Board for her excellent logistic and administrative support. Finally, I especially want to thank my fellow committee members. They worked hard, gave unselfishly of their time, and patiently learned the language and biases of different scientific disciplines while they worked to meet our charge.

                                          Michael Roman, Chair
                                          Committee to Review the Gulf of Alaska
                                          Ecosystem Monitoring Program

Acknowledgments

This report has been reviewed in draft form by individuals chosen for their diverse perspectives and technical expertise, in accordance with procedures approved by the National Research Council's Report Review Committee. The purpose of this independent review is to provide candid and critical comments that will assist the institution in making its published report as sound as possible and to ensure that the report meets institutional standards for objectivity, evidence, and responsiveness to the study charge. The review comments and draft manuscript remain confidential to protect the integrity of the deliberative process. We wish to thank the following individuals for their review of this report:

Kenneth H. Brink, Woods Hole Oceanographic Institution, Massachusetts
Ingrid C. Burke, College of Natural Resources, Fort Collins, Colorado
Robert B. Gramling, University of Southwestern Louisiana, Lafayette
Mahlon C. Kennicutt, Texas A&M University, College Station
John J. Magnuson, University of Wisconsin, Madison
Sharon L. Smith, Rosenstiel School of Marine and Atmospheric Sciences, Miami, Florida
Judith Vergun, Oregon State University, Corvallis

Although the reviewers listed above have provided many constructive comments and suggestions, they were not asked to endorse the conclusions or recommendations nor did they see the final draft of the report before its release. The review of this report was overseen by Garry Brewer of Yale University. Appointed by the National Research Council, he was responsible for making certain that an independent examination of this report was carried out in accordance with institutional procedures and that all review comments were carefully considered. Responsibility for the final content of this report rests entirely with the authoring committee and the institution.

Contents

EXECUTIVE SUMMARY 1
    Elements of a Sound Long-Term Science Plan, 3
    Conceptual Foundation, 4
    Scope and Geographic Focus, 5
    Organizational Structure, 6
    Community Involvement, 7
    Data and Information Management, 8
    Synthesis, Modeling, and Evaluation, 9
    Conclusions and Recommendations, 10

1   PLANNING LONG-TERM ECOSYSTEM SCIENCE   17
    The Committee's Charge, 19
    Elements of a Sound Long-Term Science Plan, 20

2   THE IMPORTANCE OF A CONCEPTUAL FOUNDATION   31
    The Science Plan as a Bridge Between the Conceptual
       Foundation and a Working Science Program, 34

3   DETERMINING SCOPE AND GEOGRAPHIC FOCUS   37
    Scope, 37
    Geographic Focus, 37
    Habitats as a Divisional Unit, 40
    Choice of Variables and Research Projects, 43

4	ORGANIZATIONAL STRUCTURE	52
5	COMMUNITY INVOLVEMENT AND TRADITIONAL KNOWLEDGE	59
6	DATA AND INFORMATION MANAGEMENT	66
7	SYNTHESIS, MODELING, AND EVALUATION Synthesis, 69 Modeling, 71 Review of GEM Science Background Section, 72	69
8	CONCLUSIONS AND RECOMMENDATIONS	75
	REFERENCES	84

APPENDIXES

A	Biosketches of the Committee's Members	89
B	Acronyms	93

Executive Summary

"It is a piece of ancient Greek wisdom that counting and measuring things is a much surer path to knowledge and understanding than any other." (McCready, 2001)

In March 1989 the tanker *Exxon Valdez* ran aground on Bligh Reef in Prince William Sound, Alaska, and spilled about 11 million gallons of oil. One element of various legal proceedings occurring as a result of the spill was a civil settlement that required Exxon Corporation to pay $900 million over 10 years to restore resources injured by the spill and compensate for reduced or lost services the resources provide. The *Exxon Valdez* Oil Spill Trustee Council composed of three federal and three state members was established to administer the funds. As part of its mission, the Trustee Council has disbursed substantial funding for research, first for damage assessment activities and later for monitoring and research. Significantly, the Trustees also set aside some of the funds to create a permanent trust intended to support continued, long-term research and monitoring in the region after the settlement period had ended.

Planning for this new activity, called the Gulf Ecosystem Monitoring (GEM) program, is now well under way. To help ensure that the GEM program is based on a science plan that is robust, far-reaching, and scientifically sound, the Trustee Council asked the National Academies to serve as an independent advisor. In June 2000 the National Academies appointed a special committee and charged it to review the scope and content of the program as it evolved. During the committee's two-year tenure it met multiple times with Trustee Council staff and with scientists and community members to learn about the program's intended goals and structure. To date, the committee has provided two written reports: a short letter report (November 2000) that comments on the program planning schedule and a more detailed interim report (February 2001) that critiques an early draft of the GEM program science plan (EVOSTC, 2001).

The Trustee Council is to be commended for its foresight in setting aside money over the years to create the trust fund that will provide long-term support to the GEM program. As envisioned, that program will offer an unparalleled opportunity to increase understanding of how large marine ecosystems in general, and Prince William Sound and the Gulf of Alaska in particular, function and change over time. The committee believes that this program has the potential to make substantial contributions of importance to Alaska, the nation, and environmental science.

According to an early Trustee Council document, Restoration Update Winter 2000 (EVOSTC, 2000b), GEM was conceived to have three main components: long-term ecosystem monitoring (decades in duration); short-term focused research (one to several years in length); and ongoing community involvement, including use of traditional knowledge and local stewardship. The committee views this early simple vision of the program as a sound foundation upon which to build. In a later document (EVOSTC, 2000a), the purpose of the GEM program is further delineated to contain five program goals: detect, understand, predict, inform, and solve. The committee understands the general intent of these goals and the necessity of making the program responsive to both the needs of science and the needs of various agencies and the public. Nevertheless, as the committee discussed in its interim report, it remains concerned that these five goals are extremely diverse and far-reaching. While the GEM mission is a good general statement of intent, the committee remains concerned that such broad ambition exposes the program to the risk that it will be spread too thin to be effective.

This report reviews the planning document entitled "Gulf of Alaska Ecosystem and Monitoring Program" (NRC Draft), Volumes I and II, provided in September 2001 (EVOSTC, 2001). During the course of this study, the committee saw progress in a number of areas. For example, the committee believes that the GEM planners made a significant effort to include the interests of diverse stakeholders (the Trustee Council, scientists, various advisory groups) in the science plan. We are pleased to see that the planning process has caused an evolution in the draft and the thinking behind it. We commend GEM planners for not taking the easy route of simply picking stations and starting data collection, and for taking the time to think about the conceptual foundation and develop the hypotheses that are necessary to define data needs. Finally, we find that the conceptual foundation is much improved from earlier drafts and discussions; however, placing the conceptual foundation deep within Volume II is not appropriate because this late placement implies that it is an afterthought and not the foundation upon which the program is built. We conclude that GEM planners have made progress on the development of research hypotheses, although there is still room for more work in this area.

GEM staff has made good efforts to involve the science community in its planning activities. Through these contacts they have made a solid start on plans to use modeling effectively and in developing a data management strategy. The committee found that the science review section is very useful. Although it may seem obvious, many of these positive strides have occurred because the Trustee Council and GEM staff have set up a planning process and are allowing adequate time for input, discussion, and revision. This process will make for a significantly better program over the long term.

The committee has struggled, however, with its basic charge—to review the GEM program—because the science plan was literally evolving as we worked and we often were aiming at a moving target. We also struggled because, as scientists, we are more accustomed to dealing with research programs either instigated directly by scientists, such as the Global Ecosystem Dynamics program, or by agencies with clear mandates, such as Minerals Management Service's Environmental Studies Program. Instead, GEM is a research program directed by a Trustee Council made up of six agency representatives, each carrying responsibilities for mission-oriented state and federal agencies. The Trustee Council's role is made especially difficult because of the legal requirement that all its decisions be unanimous. GEM is supported by a staff that includes both scientists and non-scientists who have the unenviable job of balancing not only the expectations of the science community (the norm when developing a new science program) but also the expectations of various other Alaskan stakeholders and the inevitable political forces present in the Trustee Council itself.

While this committee whole-heartedly endorses the idea of a long-term ecological research program in the Gulf of Alaska and commends the Trustee Council and other public decision makers for having the foresight to create such a program, we want to be clear that this report is not an endorsement of implementation of the GEM program as currently designed.

ELEMENTS OF A SOUND LONG-TERM SCIENCE PLAN

The GEM program offers an unparalleled opportunity to increase our understanding of the functioning of large marine ecosystems in general and the northern Gulf of Alaska and its adjacent waters in particular.

Few other research programs have a century-long time horizon. Thus, along with the opportunity afforded by GEM comes an obligation to craft a research plan that can endure over time. This plan requires a core set of measurements that can be taken consistently and indefinitely, as well as some flexibility to adjust to changes in conceptual understanding and research interests.

Recent research evaluating coastal monitoring studies has identified seven themes necessary in all successful programs (Weisberg et al., 2000):

1. Clearly define program goals and anticipated management products.
2. Recognize the differences between physical and biological monitoring.
3. Accommodate differences in space-time scales among ecosystems as they affect sampling design.
4. Develop an effective archival and data dissemination strategy.
5. Develop data products that will be useful to decision makers.
6. Provide for periodic program review and flexibility in program design.
7. Establish a stable funding base and management infrastructure.

The committee concurs that these broad steps are central to all good research programs. In addition, the committee has identified a number of specific elements it deems essential for a successful long-term science program of the magnitude of GEM. These include development of a clear, strong conceptual foundation for the program, early definition of a geographic scope and focus for study, an organizational structure led by a qualified chief scientist, involvement of stakeholders in the planning process and research, substantial attention to data management to ensure safekeeping and accessibility, and periodic assessment of progress through synthesis and evaluation. The committee's report is structured into sections addressing these key elements.

CONCEPTUAL FOUNDATION

The GEM program is conceived as a long-term monitoring program, because long time series are essential to detecting ecosystem change. However, it is absolutely vital to recognize that long-term monitoring per se will not necessarily lead to a better scientific understanding of the ecosystem. The value and utility of monitoring depends critically on the variables measured, the spatial and temporal extent and intensity of sampling, and the methods employed. Without a clear vision of the desired goals at the outset it is very difficult to establish monitoring programs that will provide data that will actually be useful over time. This is why the monitoring program must have a strong conceptual foundation and be driven by broad, "big-picture" hypotheses.

For GEM the conceptual foundation needs to be broad, precisely because of the long time scale of the program. No one can know which theories, taxa, or processes will emerge as critical to the public or managers, or relevant to ecosystem functioning in future decades. Conceptual foundations that rest on a few indicator species, highly specific hypotheses (e.g., Pacific Decadal Oscillation), or current human impacts (e.g., fishing) are likely to be too narrow and inflexible to support the GEM mission. Instead, GEM must incorporate the sense that marine ecosystems change in response to physical and biological changes and human impacts, as is clearly expressed in the GEM mission statement. GEM planners are aware of the difficulty of pursuing long-term monitoring in the face of short-term interests: The GEM program has provisions for multi-decade measurements and for shorter research programs targeting specific issues or hypotheses, so that GEM can respond to current concerns without sacrificing the gathering of long-term data sets that will prove increasingly useful as they accumulate.

Given its importance as a foundation and guiding force, the GEM conceptual foundation should not be hidden in Volume II of the draft science plan (EVOSTC, 2001); it should be located early in the articulation of the GEM science plan.

SCOPE AND GEOGRAPHIC FOCUS

Three important, interrelated elements must be addressed when defining the scope of a science plan, as a way of focusing attention on a practical subset of the many possible research questions. The first two elements, geographic focus and research approach, serve to set bounds on "where" the plan is applied. The geographic focus delimits the spatial extent of the plan. Research approach is the decision about how to divide research efforts in the geographic area (e.g., habitat types, species, flows of energy or materials, or the consequences of specific perturbations). The third component of scope, determining generally "what" will be measured, follows once the first two elements are agreed on and involves the selection of long-term variables to measure.

When resources are finite, there are inevitable tradeoffs between the intensity and geographic scope of research. Given finite funds, multiple variables can be monitored in a small area or fewer variables can be measured in a larger area. The choice of geographic scale for a long-term science plan is based on considerations such as scientific criteria, the existing knowledge base, management needs, accessibility, and cost.

The GEM plan has taken the entire Gulf of Alaska as its geographic scope. In its interim report the committee recommended that GEM first focus long-term research in Prince William Sound, and then extend geographic coverage over time. The rationale underlying this recommendation was the difficulty of designing a useful research plan for such a broad area given limited funds, coupled with the utility of extending existing time series at the core of the area affected by the spill in 1989. Nevertheless, the Trustee Council is well within its prerogative to select any geographic scope, but if the program is to be successful, the scope should be justified on science and management grounds and must be appropriate to the funding level. Covering a large geographic scope in the absence of a scientific rationale (a unifying hypothesis) risks expending resources in a piecemeal fashion that will make synthesis and interpretation difficult.

Because of the tradeoff between geographic scope and intensity of research effort, science plans covering large areas must include methods for stratifying observations and allocating funds. This focus can be provided in a number of ways, including an emphasis on habitats (as selected by GEM planners) or with other organizing concepts such as species, hypotheses, time, or flows of energy. In the GEM planning document (EVOSTC, 2001), the decision to organize by habitat is acceptable, but there are several problems that should be addressed. In the draft plan, hypotheses are presented as repetitive questions in each habitat type, and they will need considerable refinement before they can guide research. Most importantly, the habitat divisions may create a barrier to understanding links and transfers among habitats. The committee cautions against the development of habitat-based subcommittees in the organizational structure, as there is substantial risk of neglecting linkages among habitats.

Different strategies will be required for the three types of research included in the GEM plan—measuring variables long-term, carrying out shorter-term studies of processes, and synthesizing and analyzing collected data sets. It is appropriate to devote considerable time and effort to making effective choices of what, where, and when to measure. The committee finds little indication that hypothesis testing will play a role in designing long-term research. Without clear hypotheses, there is little guidance on how these variables will be chosen, although the process appears to include some modeling, gap analysis, and workshops.

ORGANIZATIONAL STRUCTURE

A credible scientific program must assure that the science base is sound and that program planning, implementation, community involvement, coordination, proposal solicitation, peer review, funding, interactions among investigators, data management, program oversight and review, and public outreach are efficient. Most interdisciplinary marine ecosystem programs have a scientific steering committee (the equivalent of the Scientific and Technical Committee proposed by GEM planners [shown in Figure 4-1]) and a chief scientist or scientific director that together develop and implement the science plan and provide program oversight. The chief scientist works closely with the steering committee, but is ultimately responsible for developing and implementing the program science plan, and has authority regarding all scientific decisions after consultation with the principal investigators and steering committee. The GEM plan does not include detail on organizational structure, but a flowchart provided by staff (Figure 4-1) contains the necessary elements, although how these elements are implemented and given authority for real action is, of course, key.

Science planning must continue during the life of the GEM program to assure program success. The core variables to be measured must be carefully selected and should not be modified without careful consideration during the life of GEM. This strategy will assure that consistent long-term data are obtained with the principal objective of distinguishing between human induced and natural changes in the Gulf of Alaska ecosystem. The Scientific and Technical Advisory Committee may be of value in both developing monitoring protocols and requests for proposals, but such a committee should not be the sole mechanism by which the variables to be measured are selected. Other input might be sought through targeted workshops designed to synthesize existing knowledge and determine the location and frequency of measurements of key biological, chemical, and physical variables.

COMMUNITY INVOLVEMENT

Community involvement and the incorporation of traditional knowledge is critical to the GEM program's long-term success. Early GEM-related documents indicated a clear desire to incorporate community involvement and traditional knowledge, however this emphasis appears to have receded in successive documents. The committee urges the Trustee Council to reconsider this change in emphasis.

Why is incorporation of community involvement and traditional knowledge important? First, community involvement and traditional knowledge can contribute to the overall focus on ecosystem monitoring. Local residents possess valuable ecological knowledge that can be directly incorporated into established scientific models. Local residents can be a source of important research questions and can help assure that research is relevant to both ecological and community needs. In addition, local residents offer potential efficiencies in data collection efforts.

A second rationale relates to equity issues. The GEM program, like the Trustee Council itself, is the result of settlement funds dedicated to restoration of an ecosystem damaged by a human technological disaster (Erikson, 1994). This damaged ecosystem includes resource-dependent human communities (Picou and Gill, 1996), and these stakeholders have a justifiable interest in the outcome of the resulting activities.

Public review does not equal public involvement, although it should be part of an overall commitment to public involvement. Meaningful community participation must consist of more than providing employment to local residents (to work on projects conceived and run by others), and must include participation in developing the actual research questions. This does not mean that employing local residents is inappropriate but rather that the continued identification of involvement exclusively with employment is unnecessarily narrow.

Community involvement should be designed to promote meaningful participation and provide for flexibility as the GEM program evolves. In many respects the program will be breaking new ground in terms of integrating community involvement into a long-term science plan. The committee is under no illusion that successful incorporation of community involvement and traditional knowledge in the program will be easy, but we conclude that it is necessary.

DATA AND INFORMATION MANAGEMENT

The legacy of the GEM program will be the data it collects. Given the objective of establishing a long-term measurement program in the Gulf of Alaska and its importance to both regional and national interests, GEM must make a strong commitment to data and information management. The goals must be to facilitate data exchange among GEM scientific investigators, make data available to the public and others outside the scientific community, and archive GEM data products.

GEM will need to make a major commitment to fund data management activities, probably through a Data Management Office composed of a data manager, assistants, and the necessary infrastructure to organize, disseminate, and archive data. That office would develop data policies, implement a data management system, ensure preservation of data with relevant documentation and metadata, review data management efforts, enforce data policies, and facilitate exchange of data with related oceanographic programs. GEM needs to be committed to the timely submission and sharing of all data collected by its researchers.

Data management must have sufficient resources to accomplish its mission. Successful coastal monitoring efforts allocate as much as 20 percent of their total budget to data management (Sustainable Biosphere Initiative, 1996; Weisberg et al., 2000).

The general description of the data management architecture in the draft GEM science plan is very good. The basic functions of data receipt, quality control, storage and maintenance, archiving, and retrieval are adequately addressed. The report recognizes that different types of data products will be needed for basic research and analysis, modeling, resource management applications, and public outreach. Access to the data archives and software display will be an important component of public outreach. There will be multiple levels of complexity to data access, ranging from users with limited experience to use by the investigators who gathered the data.

SYNTHESIS, MODELING, AND EVALUATION

The committee understands the difficulty of writing a science plan to guide the GEM program for the next 100 years. It is simply not possible to know everything that should be addressed. Thus, the plan will need to be flexible. It must include procedures requiring synthesis of knowledge at specific points in time and opportunities to evaluate past efforts and make adjustments in direction.

An initial synthesis needs to include several components. The first step for the GEM program to be successful, a much needed literature review, has been completed in the "Scientific Background" section in Volume II, Part 3 of the GEM plan. The second step, compilation, assessment, and analysis of data, has not been done. This step is critical to the third step, which is a synthesis of *Exxon Valdez* oil spill research from 1989 to the present. Although a few Trustee Council-supported programs have completed synthetic views of their results (e.g., *Fisheries Oceanography*, Vol. 10, Suppl. 1, "A Sound Ecosystem Assessment Synthesis"), many have not.

The knowledge gained and publicized about Prince William Sound is extensive because of Trustee Council funding. Retrospective analyses have led to new hypotheses and ideas in many instances; there is, however, much more to be gained from the past studies that should be used to direct the future of GEM. The synthesis of data and assessment of what has been learned in the recent studies will provide a baseline from which to develop hypotheses to guide GEM research. Annual reports are not peer-reviewed publications and do not qualify as syntheses.

Synthesis and modeling are interconnected. For example, initially one could create a conceptual model to identify quantities that need to be measured, collect data, synthesize data, and then create a more refined quantitative model. Alternatively, one could collect and synthesize data, and then generate a statistical model that could be used to collect more data to verify the model. Regardless of the order of these steps and the sophistication of the techniques, the components of synthesis and modeling are both critical. The combination of synthesis and modeling provides tools for evaluation of past work, testing the appropriateness and accuracy of hypotheses, and generation of new hypotheses.

The elements of a successful modeling component are outlined in the GEM plan. The GEM program should work toward more realistic and accurate numerical models for the prediction of ecological processes. The unparalleled opportunity of a long-term observation program in the Gulf of Alaska, coupled with a concerted effort in modeling, will produce exciting new tools for the management of the Gulf of Alaska's ecological resources.

CONCLUSIONS AND RECOMMENDATIONS

Opportunity for Sustained Study

Conclusion: GEM is an important opportunity to do truly long-term research in a marine ecosystem, and this long-term approach is essential to distinguish natural variability from human impacts. The long-term nature of the program, intended to cover a period of many decades, is the flagship contribution of the plan. Long-term monitoring by definition must include sustained, consistent observations over a long period and thus requires a long-term commitment from the highest levels of decision makers. This commitment will require a substantial financial investment. Short- and medium-term research is an appropriate way to address current questions and management needs, but the fundamental importance of the long-term program should not be lost.

Recommendation: The majority of GEM funds should be spent on long-term monitoring and research, that is, sustained observations of ecosystem components and ecological processes over decades. This long-term perspective will be the GEM program's special contribution to scientific understanding in Alaska's marine environment; most other research programs are short-term. These long-term measurements will be necessary to differentiate the effects of natural variation from human-induced changes on the Gulf of Alaska ecosystem. The coastal Long-Term Ecological Research sites funded by the National Science Foundation provide good models of such long-term research.

Elements of a Sound Long-Term Research Plan

Conclusion: A sound, long-term research plan must clearly define its conceptual foundation, scope, organizational structure, data management methods, and methods for periodic synthesis and review. The conceptual foundation presented in the draft science plan is adequate and with modest restatement as a hypothesis could be a useful focus for research. The science plan and research objectives need to be directly linked to this conceptual foundation.

Recommendation: The current draft science plan (EVOSTC, 2001) needs to be shortened considerably by removing tangential materials so that it is a clear guide for the future. The conceptual foundation needs to be discussed early in the GEM planning document because that placement captures its importance as the fundamental building block on which the rest of the program depends. The science plan should include a broad conceptual foundation that is ecosystem-based. It should seek to understand natural and human-induced changes and it should be flexible to accommodate changing needs without compromising core long-term measurements. These hypotheses will provide a bridge between the conceptual foundation and the eventual implementation of the science program. Because the conceptual foundation states that the ecosystem is affected by both natural variability and human-induced change, as the plan is implemented both of these drivers should be addressed in studies.

Implementation of the GEM Program

Conclusion: The planning process for GEM has been difficult and costly, but the investment in planning is critical for success. Long-term measurements cannot begin until after the appropriate variables have been identified, and these must be based on the conceptual foundation and hypotheses. The planning and design of sampling will continue to take considerable time and effort in the early years of the program. It is more important to identify the right variables than to rush to collect data.

Recommendation: The GEM plan and planning process need to provide careful consideration of what to measure, how often, and where, based on input from a broad cross-section of the scientific community, local communities, and managers. These decisions on hypotheses and attendant measurements should be made by the chief scientist working with the Scientific and Technical Advisory Committee and other independent scientists and stakeholders over the course of several years as program implementation gets under way.

GEM's Role in Gulf of Alaska Research

Conclusion: GEM's primary goal should be to develop a comprehensive and eventually predictive understanding of the Gulf of Alaska ecosystem. The long-term nature of GEM will enable it to serve as a framework for marine research in the Gulf of Alaska. Other programs will come and go on shorter time frames and should be encouraged to coordinate with GEM, but GEM does not have the resources to be the central coordinating body for all such efforts.

Recommendation: The focus of GEM should be its long-term program, and GEM decision makers should not try to do too much or this will dilute GEM's limited resources and impact. Because of the long time frame of GEM, it can provide a building block for partnering with other programs that will come and go, but it should not be distracted by the idea of assuming leadership of Gulf of Alaska marine research.

Recommendation: GEM should not see its role as filling the gaps in other programs, because adding these kinds of activities will inevitably erode funding for the GEM core measurements. This does not preclude GEM from involvement in other programs in which the research is addressing issues or collecting data that has been identified as necessary for addressing the central hypotheses of GEM.

Recommendation: It simply is not possible for GEM, given its resources, to play a leadership role in both scientific research and day-to-day support of resource management. GEM should not be involved in the types of monitoring that are typically the responsibilities of agencies. GEM should not subsume routine surveys, stock assessments, and data collection that have been the normal province of resource management agencies. Of course, a large monitoring program like GEM will supply much information that is useful to resource management agencies as a result of its own activities.

Community Involvement

Conclusion: The GEM plan does not currently describe effective and meaningful ways to involve local communities. This involvement should occur at all stages, from planning (e.g., selecting the questions to be addressed and variables to be monitored) to oversight and review. Local knowledge and traditional ecological knowledge can be used to generate ecologically sound and socially relevant research ideas. Science and community partnerships can lead to achievements that neither could attain independently. Specifically, such collaborations provide scientific knowledge as well as community education and local support of science. These outcomes are important especially because of the long-term nature of GEM; such involvement might be less critical in shorter programs, but the century scale requires the establishment of long-term bonds.

Recommendation: The Trustee Council and GEM program staff must continue to seek ways to build meaningful community involvement at all stages of planning and implementation, from selecting the questions to be addressed and identifying the variables to be monitored to providing program oversight. It was outside the scope of this committee to advise specifically on what programs or methods to use; neither are we as experienced as GEM staff in dealing with Alaska's diverse communities of interest. Nonetheless, we are certain that the community involvement debate will continue until better resolution of this issue is found.

Geographic Scope

Conclusion: No program can be expected to meet the needs of all potential data users, and tradeoffs are inevitable between the intensity and spatial range of sampling. That is, if the scope of GEM is physically large, then its long-term research component will be able to collect less information at any one site (because there is a finite amount of information that can be collected with finite financial resources). If the scope of GEM is physically smaller, there can be more monitoring sites or more types of information collected. Research projects and sampling will need to be selected very carefully to avoid diluting activities so that their usefulness is limited. GEM planners can choose to obtain more limited information from a large area or more in-depth information from a smaller area.

Recommendation: GEM planners must make an explicit choice on how to focus the program's research. There are many options for carrying out coordinated research that avoid piecemeal projects. One option is to concentrate on a particular geographic area, as the committee recommended in its interim report. Another possibility is to target a few variables across a broad geographic range, such as measuring physical oceanographic variables over long time periods (e.g., temperature, salinity, currents). It is possible to concentrate attention on particular habitats in a large geographic range. These choices must be guided by the conceptual foundation and the hypotheses selected for investigation.

Using Habitat as an Organizing Concept

Conclusion: GEM or any large research program can organize its effort and funds in many ways and still be successful. The habitat approach described in the GEM science plan is one way of dividing attention and funds, and it has the advantage of being understandable to many of the program's key stakeholders. GEM planners need to be aware of its one critical disadvantage: A habitat approach can fail to address key linkages, flows, and processes between habitats, which is where many of the most interesting lessons of the long-term GEM program might be seen.

Recommendation: Given the habitat approach selected, GEM planners must make a concerted effort to ensure that the program has clear, concrete mechanisms to address cross-habitat links. This does not necessarily mean creating a linkage subcommittee but rather building into each habitat study the opportunity to make measurements of flows among habitats and highlight other interactions. Across-habitat connections must be addressed during synthesis and modeling. These efforts are essential to creating a truly integrated program, where the whole is greater than the sum of the parts.

Organizational Structure

Conclusion: The GEM research plan is being developed to carry out long-term research, short-term research, and synthesis and modeling of data sets. Soliciting proposals, evaluating proposals, and the time frame for the research effort and its funding will differ for these scientific activities. The current science plan does not distinguish these activities in terms of the procedures necessary to manage them and achieve useful results, or even that the goals of these three approaches differ. Strong scientific guidance is required through all the activities of GEM.

Recommendation: GEM planners, with input from the science community, should identify how these three kinds of scientific endeavors will be incorporated and managed within the science plan. For instance, long-term research projects, short-term research projects, and synthesis efforts will require different mechanisms for proposal solicitation and evaluation and different time frames for funding.

Recommendation: The scientific leadership of the GEM program should be in the hands of a chief scientist advised by a Scientific and Technical Advisory Committee. The chief scientist should have adequate assistance to execute the program.

Conclusion: The organizational structure supporting GEM needs to ensure ongoing, independent scientific oversight and review. It should be easy for new researchers and local community members to be involved in planning and carrying out the research projects. If the Scientific and Technical Advisory Committee is to function effectively and play a leadership role in developing and directing the GEM scientific and technical program, its membership must be selected carefully.

Recommendation: The Scientific and Technical Advisory Committee will play a key role in leading the GEM program and ensuring program credibility. Committee members should be chosen based on their scientific expertise and their ability to link across the marine habitats and disciplines. To obtain the best program oversight over time there should be regular rotation of the members of all advisory groups, such as the Scientific and Technical Advisory Committee. Advisory Committee members should be, and should be perceived to be, neutral parties who are focused on the long-term success of the program. Members may need to be compensated for their service; they should have term limits of three to five years with no direct GEM research funding during their period of service.

Recommendation: The design of proposal solicitations and final recommendations for Trustee Council funding should be major functions of the Scientific and Technical Advisory Committee and chief scientist. In designing proposal solicitations, the Advisory Committee should be responsible for developing the scientific and technical subjects required to address GEM goals. Community workshops hosted by the Scientific and Technical Advisory Committee would be one method to help articulate community-generated research needs and could be a way to increase the participation of local communities that use Gulf of Alaska resources. The Scientific and Technical Advisory Committee and chief scientist should be responsible for organizing workshops designed to provide input on core variables to be measured over time. Final decisions on variable selection can be based on hypotheses proposing how each variable provides insight into human and climate-based changes in the ecosystem.

Recommendation: There should be an open process for nominating individuals to serve on the Scientific and Technical Advisory Committee, both during its initial formation and as the GEM program continues. Various independent scientific groups can assist in the initial formation to help broaden the selection process and find candidates with suitable experience in the initiation and implementation of large-scale, long-term ecological research. The chief scientist should review the nominations and recommend selections with appropriate documentation to the Trustees, who are responsible for the appointments.

Data and Information Management

Conclusion: There will be significant costs associated with data and sample processing and with data archiving. It is a common mistake to underestimate the cost of data and information management. To extract the full scientific value of any research program, data and information must be made available to the scientific community, resource managers, policy makers, and the public on a timely basis. Each of these audiences will require information in a different format. The committee commends the initial development of data management procedures; careful implementation of these procedures is key.

Recommendation: GEM should create a comprehensive Data Management Office (not just an archive but a group of people who address these issues). Other large science programs spend as much as 20 percent of funds on data management. The multi-decadal scale of GEM will require a similar commitment.

1

Planning Long-Term Ecosystem Science

In 1989 the T/V *Exxon Valdez* spilled about 11 million gallons of crude oil into Prince William Sound in Alaska, setting off a cascade of effects that still have repercussions more than a decade later (Figure 1-1). One result was that in 1991 the U.S. District Court approved a civil settlement that required Exxon Corporation to pay the United States and the State of Alaska $900 million over 10 years to restore the resources injured by the spill and to compensate for the reduced or lost services (human uses) the resources provided. Under the court-approved terms of the settlement the *Exxon Valdez* Oil Spill Trustee Council made up of three federal and three state members was formed to administer these funds. The mission of the Trustee Council has been to return the environment to a "healthy, productive, world-renowned ecosystem" by restoring, replacing, enhancing, or acquiring the equivalent of natural resources injured by the spill and the services provided by those resources. It also set aside some of the funds to create a permanent trust to support continued, long-term research and monitoring in the region. At this point the Trustee Council is developing a plan to guide this new research program, to be known as the Gulf Ecosystem Monitoring (GEM) program.

As part of its mission the Trustee Council has disbursed research funds for almost 10 years, at first for damage assessment activities and then for monitoring and research to better understand the ecosystem and to understand impacts of the oil spill on important "resource clusters," or communities/resources (e.g., salmon, herring, marine mammals, subsistence resources). Extensive research has been conducted over the decade, making this the most studied cold water marine oil spill in history. In

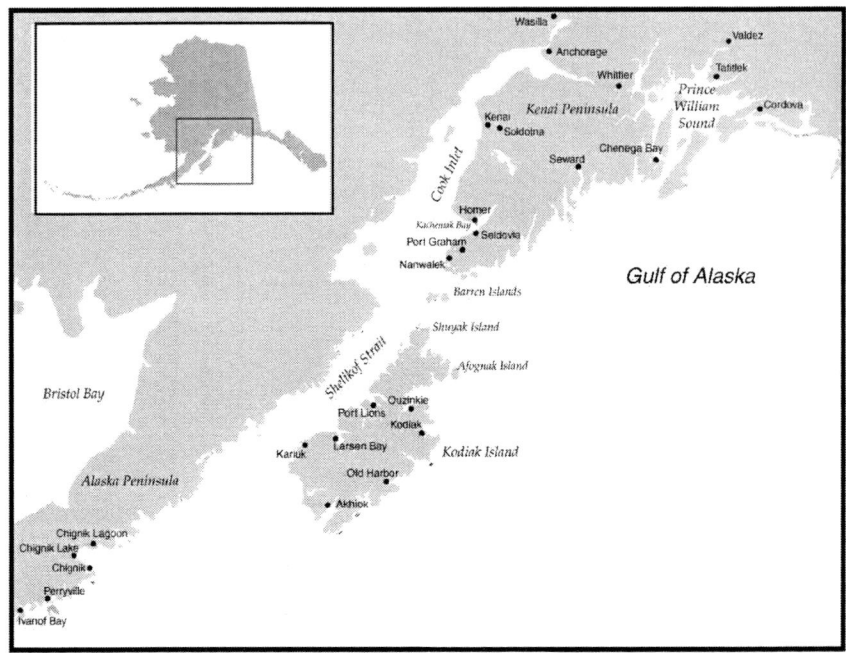

FIGURE 1-1 Region encompassed by GEM.

keeping with its mandate and after extensive public input the Trustee Council decided to use the trust fund to support continued research and monitoring in the region into the future. The GEM program has a unique opportunity to obtain the long time series of data necessary to support research on the effects of decadal-scale change on the structure, function, and ability of a marine ecosystem to provide goods and services to people. This research program will provide the depth and continuity of data collection necessary for both practical management lessons and deeper understanding of the causes and effects of ecosystem change.

The Trustee Council showed great foresight in setting aside funds over the years to create the trust fund that will now provide long-term funding to the GEM program. As envisioned, the program will offer an unparalleled opportunity to increase understanding of how large marine ecosystems in general, and Prince William Sound and the Gulf of Alaska in particular, function and change over time. The committee believes that it stands to be a significant program of importance to Alaska, the nation, and the scientific community.

THE COMMITTEE'S CHARGE

To ensure that its plan for long-term research and monitoring in the Gulf of Alaska ecosystem is the best possible, the Trustee Council asked the National Academies for assistance, and a specially appointed committee was formed to review the scope, content, and structure of the draft science program and draft research and monitoring plan. The Committee to Review the Gulf of Alaska Ecosystem Monitoring Program was asked to provide independent scientific guidance to the Trustee Council, research community, and public as the Trustee Council develops a comprehensive plan for a long-term, interdisciplinary research and monitoring program in the northern Gulf of Alaska. Specifically, the committee was charged to:

- gain, through briefings and literature review, familiarity with the relevant body of scientific knowledge, including but not limited to that developed by the research and monitoring activities sponsored by the Trustee Council in the past.
- convene one or more information-gathering meetings in Alaska, where researchers, the public, and other interested people can convey their perspectives on what the research and monitoring plan should accomplish.
- review the general strategy proposed in the draft science program (which includes information on the social and political context, mission, approach, and scientific background) and make suggestions for improvement.
- review the draft research and monitoring plan, including the scope, structure, and quality of the approach proposed for a long-term research and monitoring program in the northern Gulf of Alaska. This includes whether the conceptual foundation provides an adequate basis for long-term research and monitoring, and whether the research and monitoring plan adequately addresses gaps in the knowledge base and existing uncertainties.

Since this committee was formed in June 2000 we met five times to discuss the GEM program and consider the strengths and weaknesses of the program's planning documents. We have conveyed our comments and recommendations in a letter report (November 2000) with advice on program timing and in a more detailed interim report (February 2001) that critiqued an early draft of the program's science plan. These reports focused on the early planning for GEM, were specific to the draft planning documents, and were primarily directed to program staff. In this final report we provide broader comments and a document that has more general and longer-lasting lessons about which elements are essential to the success of a long-term research and environmental monitoring program such as GEM.

ELEMENTS OF A SOUND LONG-TERM SCIENCE PLAN

The world's oceans have long been viewed as producing an inexhaustible supply of protein and other goods and services for human use. But evidence of the adverse effects of human activities on marine ecosystems is increasing and reminding us that the ocean's resources are not inexhaustible (NRC, 1999a). It is increasingly clear that the structure and functioning of marine ecosystems is profoundly linked to variability and changes in ocean climate and that those changes can occur rapidly. One of the greatest challenges facing society, and particularly managers of marine living resources in the Gulf of Alaska and elsewhere, is to understand the relative effects of human activities and natural changes in ocean climate on the goods and services supplied by marine ecosystems (NRC, 1996).

Why is this so difficult? One reason is that marine ecosystems are large, complex interactive systems in which organisms, habitats, and external influences act together to regulate both the abundance and distribution of species (NRC, 1999a). Species interactions and the effects of variability in ocean climate on those interactions occur at spatial scales ranging from centimeters to hundreds of kilometers and on temporal scales ranging from minutes to decades. Human activities also act at various scales and may act selectively on certain components of an ecosystem (e.g., higher trophic levels), although such activities can have cascading effects throughout marine ecosystems (Carpenter et al., 1985; NRC, 1996). These disparate spatial and temporal scales make it difficult to measure the processes affecting marine ecosystems and to monitor ecosystem structure and functioning (Weisberg et al., 2000). The diversity of temporal scales at which important processes affect marine ecosystems makes it difficult to measure many of these processes over short periods of time. Finally, perturbations to marine ecosystems often appear to act in subtle, nonlinear ways making it difficult to understand the consequences on ecosystem components that may be of particular interest to society, such as birds, mammals, and fishes. Given these challenges, we commend the Trustee Council for having the vision to develop a long-term ecological monitoring program that stands to have great enduring value to the stakeholders of this vast and diverse marine ecosystem.

Good management requires good information and the knowledge of how to use this information to predict the outcome of management decisions. Thus, a prerequisite of good management is good science. As the committee noted in its interim report, given the complexity of marine ecosystems and the failure of single-species management to produce sustainable fisheries in many parts of the world (NRC, 1999a), it is not surprising that both scientists and managers have increasingly promoted the concepts of multi-species or ecosystem-based management. However, it is clear that not enough is known about most large marine ecosystems, including the Gulf of Alaska, to implement a useful whole-system approach to management.

It is reasonable to ask what an ecosystem-based approach to management could provide in the medium term that a single-species approach cannot. The National Research Council's Committee on Ecosystem Management for Sustainable Marine Fisheries considered two benefits (NRC, 1999a). One benefit is that it broadens the policy framework to include a wide range of ecosystem goods and services and it acknowledges the critical role of ecosystem processing in providing those goods and services. Another benefit is that there is an explicit recognition that segments of society may have different goals and values with respect to marine ecosystems and that those goals and values may conflict. The committee believes that the promise of an ecosystem-based approach to resource management, which recognizes the changing nature of both the physical environment and species interactions and the fact that many of these changes occur at time scales greater than several years, provides a forceful scientific rationale or conceptual foundation for the GEM program. The other benefit is an explicit recognition that segments of society may have different goals and values concerning marine ecosystems and that those goals and values may conflict. To meet its goals effectively the GEM program must take a longer (interdecadal) view at appropriate spatial scales.

GEM can respond to current concerns without sacrificing long-term data sets that will prove increasingly useful as they accumulate. A well-designed and broad-based program will provide the best possible scientific basis for dealing with short-term ecological issues of public concern. Indeed, a strongly designed program will provide a sound basis for additional attention to be paid to matters of urgency or immediate public concern, even if they are not central to the program itself. However, GEM will have to be carefully constructed to avoid being excessively distracted by real or perceived ecological crises. It will, therefore, be important to define clearly not only the program goals in terms of scientific questions but also the products of the program that are expected to be of value to managers (Weisberg et al., 2000). As stated by Weisberg et al., "The most successful programs have been those with clearly defined users for the data they produce, which requires early interaction between scientists responsible for designing the program and targeted data users." The GEM program should not be used to substitute for routine monitoring and stock assessment activities that have customarily been the province of state and federal agencies. Such a use of GEM funding would constitute a tragic waste of an extraordinary opportunity.

As conceived, GEM is meant to be a long-term monitoring activity, and long time series are essential to detecting change on intermediate and long time scales. It is vital to recognize that long-term monitoring per se will not necessarily lead to a better scientific understanding of the ecosystem. The value and utility of monitoring critically depends on the variables measured, the spatial and temporal extent, and intensity of sampling. Without clear vision at the outset it is difficult to establish monitoring programs that will provide useful data for sound resource management. This is why the monitoring program must have a strong conceptual foundation and be hypothesis-driven (Box 1-1).

BOX 1-1
Providing Focus by Selecting Key Research Questions

GEM is a unique opportunity to establish a realistic long-term monitoring program. Thus one logical approach would be to shape the program around long-term monitoring as the core activity, with smaller elements added to meet other goals, and base the science plan on this two-prong structure. To make success more likely program planners would need to select a few key questions to guide the work, and these questions in turn should be based on some clear conceptual model (e.g., NRC, 1995, 2000). One way to begin is to ask what parameters are most able to provide insight into the desired questions if there is a long time series of data available. Another approach is to identify the questions for their own sake and let them suggest the parameters to be monitored.

The questions listed in Appendix C-2 of EVOSTC (2000a) are a good start. The quality and relevance of the questions suggested by members of various communities that made presentations in Anchorage on October 6, 2000, were excellent. For example, the question about the degree to which ocean conditions (productivity) affect the growth and survival of juvenile salmon and hence the degree to which science can help predict the probable percentage of returns from hatchery releases is very relevant. To answer this question requires information on physical, chemical, and biological features of the ocean, including information about salmon. Long time series of information on such factors would not only help answer the specific question but would be of great use for understanding related questions, such as insights into fluctuations in the populations of other impor-

The unique aspect of GEM is the guarantee of funding over a long time frame and the possibility of consistent, long-term measurement of species and processes in the Gulf of Alaska and Prince William Sound. Although it will require sustained commitment, long-term monitoring is an essential underpinning of the major goals of the GEM program, which stands to have great value as a model for how to monitor and understand other complex marine ecosystems. After all, the management issues facing users of Prince William Sound and the Gulf of Alaska are much the same as those found elsewhere in Alaska's marine waters and around the globe. Making long-term research the focus of GEM will create greater benefits to both basic understanding of the gulf ecosystem and its long-term management than would an abundance of short-term projects, many of which could be funded in other ways.

---

tant ecosystem components, including marine mammals, crabs, marine birds, and herring.

Several approaches could provide greater focus on GEM during implementation, even given its broad mission and goals. The committee is not recommending these as the "right" tasks, but as illustrations of the range of thinking that is necessary.

- Develop a whole-ecosystem fishery model as a guide to think about what needs to be monitored. Such a model would use current and historical data to relate yields to climate data and contaminant levels and might stress biological and physical endpoints (zooplankton and phytoplankton blooms, macrofauna populations) and climate and physical oceanography endpoints, in conjunction with modeling.

- Identify indicator taxa for monitoring. Species should be selected based on the ability of monitoring information to provide information on ecosystem functioning, not solely to reflect economic value or political importance. This takes smart choices so that the indicator species reflect a wide set of variables for measurement and serve as sentinels to provide clear and early warning of change.

- Conduct or take advantage of large-scale adaptive management studies that others implement. The Trustee Council does not have the authority to impose management changes, but it could, for example, follow population trajectories in areas with and without fishery closures or record biogeochemical variables in bays before and after aquaculture operations are instituted.

Monitoring over extremely long time periods, such as envisioned in GEM, cannot be differentiated from research; research designed to evaluate the ecological impact of climate change is of longer duration than the familiar three- to five-year process studies (Box 1-2). The development of long time series measurement is a crucial research tool for understanding ecosystem function. Along with the opportunity afforded by GEM comes an obligation to craft a research plan that can withstand the test of time. This requires a core set of measurements that can be taken consistently and indefinitely, as well as flexibility to alter both conceptual understanding and research interests. Long-term programs should be modified only when a compelling case is made that change will improve the program (Weisberg et al., 2000).

The committee identified a number of elements deemed essential for a successful long-term science program of the magnitude necessary to fulfill the mission statement and goals articulated for the GEM program by the Trustee Council (EVOSTC, 2000a). These elements are similar to those in a recent synthesis of lessons learned in a number of large-scale coastal

---

BOX 1-2
Monitoring versus Research

In oceanography today, repeated measurements made for long periods of time are typically called monitoring, while repeated measurements made over shorter periods of time are likely to be called research. While there can be other differences between monitoring and research, often the only difference between the two is the duration of the sampling. When the purpose of long- and short-term measurements is the same, that is, observing the oceans and interpreting trends, both really are aspects of scientific research. Thus, in many (if not most) cases, monitoring might just as appropriately be called research. This clarification is important only because at times the scientific community deems monitoring less meaningful than research. But for GEM, this is clearly not the case: Long-term monitoring should be the heart of the program.

Over the course of GEM, it is expected that some measurements will be made over the entire duration of the program, whereas others will be of briefer duration—years, months, days, or hours. Both timeframes of observation are important. The short-term measurements will allow the study of short-term processes, but their contributions to scientific research are not necessarily greater or lesser than the sustained observations. Indeed, a strength of the GEM program will be that it provides ocean observations of various durations with short-term sampling embedded within the sustained observations.

> **BOX 1-3**
> **Themes Needed in All Coastal Monitoring Programs**
>
> 1. Clearly define program goals and anticipated management products.
> 2. Recognize the differences between physical and biological monitoring.
> 3. Differences in space-time scales among ecosystems affect sampling design.
> 4. Develop an effective data dissemination strategy.
> 5. Develop data products that will be useful to decision makers.
> 6. Provide for periodic program review and flexibility in program design.
> 7. Establish a stable funding base and management infrastructure.

monitoring efforts (Box 1-3; Weisberg et al., 2000). In addition, the committee examined a number of existing science plans for lessons to help guide GEM planning (Box 1-4); although great variety was found in these plans, they generally confirm the importance of the elements determined by this committee as important.

Elements seen as essential to the GEM program include:

1. *A conceptual foundation.* A conceptual foundation expresses the main focus of a plan and provides a general picture of how parts of the ecosystem function and interact. A broad conceptual foundation with a sound scientific basis provides a strong scientific justification for a program and helps to defend it from criticism and political pressures over time. It provides an intellectual structure that can guide modification of the program if that becomes necessary.

2. *A scope and geographic focus for study.* In any ecosystem study, a trade-off exists between the extent of the region to be studied and the quality, density, and frequency of measurements (Weisberg et al., 2000). It is necessary to identify that portion of an ecosystem that can be monitored with sufficient intensity to provide the density of measurements needed to identify change at the desired level of scientific confidence. The *Exxon Valdez* oil spill affected Prince William Sound, the northern and western Gulf of Alaska, and lower Cook Inlet. Selecting an appropriate subset of the northern Gulf and its adjacent waters that can be studied over the long term as a connected whole will challenge the GEM program.

3. *Scientific leadership.* GEM must have strong scientific leadership. A

BOX 1-4
Common Elements of Other Science Plans

The term "science plan" has an elusive definition, encompassing documents as disparate as specific research proposed for the upcoming field season (e.g., Palmer Station Long-term Ecological Research) and new visions of multi-disciplinary research to inspire funding (e.g., RIDGE 2000). We examined a number of science plans in an effort to define our expectations of the GEM program plan. These plans are described briefly here.

1. The Long-Term Ecological Research (LTER) funded by the National Science Foundation (NSF) is perhaps the premier long-term scientific monitoring program in the United States. The coastal LTERs (e.g., Everglades, Georgia, Santa Barbara) are of particular relevance to the GEM program because they—like GEM—consider the connection between marine and terrestrial ecosystems. In addition to perhaps providing some ideas to follow as models for GEM, there are opportunities for scientific exchange between scientists working on those LTERs and GEM scientists, and perhaps even the possibility of joint activities, especially where large-scale processes are involved. Many of the LTER sites include science plans or proposals outlining the goals of ongoing research and organizational structure of personnel involved in projects and administration. <http://lternet.edu>.

2. SOLAS (Surface Ocean Lower Atmosphere Study) seeks "to achieve quantitative understanding of the key biogeochemical-physical interactions and feedbacks between the ocean and the atmosphere, and how this coupled system affects and is affected by climate and environmental change." SOLAS has three foci: (1) biogeochemical interactions and feedbacks between ocean and atmosphere; (2) exchange processes at the air-sea interface and the role of transport and transformation in the atmospheric and oceanic boundary layers; and (3) air-sea flux of CO^2 and other long-lived radiatively active gases. The science plan addresses the importance of modeling and long time series. <http://www.ifm.uni-kiel.de/ch/solas/plan-index.html>.

3. The science plan for EOS (Earth Observing System) justifies measurements being taken using a variety of remote-sensing techniques. Among science plans it is unusual in being exceptionally long (the summary alone is 64 pages), and incorporating mostly background rather than unanswered questions. No organizational structure is outlined, presumably because this fits within NASA structures: "The Earth Observing System (EOS) Science Plan is the product of leading scientists around the world who are participating in NASA's ESE/EOS program. The purpose of the Plan is to state the concerns and problems facing Earth Science today, and to indicate contri-

butions that will be made toward providing solutions to those problems, primarily through the use of satellite-based observations that will be obtained with EOS satellites and instruments." Seven focal areas are: atmospheric circulation, ocean, atmospheric chemistry, hydrology, cryosphere, stratosphere, and volcanoes.
<http://eospso.gsfc.nasa.gov/sci_plan/chapters.html>.

4. The SALSA (Semi-Arid Land-Surface-Atmosphere program) science plan was prepared by the U.S. Department of Agriculture's Agricultural Research Service to inspire and encourage collaboration. Much like the GEM program, "the Semi-Arid Land-Surface-Atmosphere Program is a multi-agency, multi-national global-change research effort that seeks to evaluate the consequences of natural and human-induced environmental change in semi-arid regions. The ultimate goal of SALSA is to advance scientific understanding of the semi-arid portion of the hydrosphere-biosphere interface in order to provide reliable information for environmental decisionmaking. SALSA will accomplish this through a long-term, integrated program of observation, process research, modeling, assessment, and information management, using both existing and innovative technologies, and sustained by cooperation among scientists and information users." Unlike the GEM program, SALSA has no money of its own: Government agencies intend to provide data management capacity and to encourage and enhance scientific collaboration. <www.tucson.ars.ag.gov/salsa/archive/documents/plans/salsascienceplan.PDF>.

5. PSAMP (Puget Sound Ambient Monitoring Program) documents are not billed as a science plan, but they demonstrate how one group has justified the use of indicators in a marine system. "Monitoring and research are vital to understanding the status of Puget Sound's health. The Puget Sound Ambient Monitoring Program (PSAMP) brings together local, state, and federal agencies—coordinated by the Action Team—to assess trends in environmental quality in the Sound. Information from the program is used to evaluate the effectiveness of the management plan and set priorities for the work plan. Through PSAMP studies, data on marine and fresh waters, fish, sediments and shellfish in Puget Sound have been collected since 1989; surveys of nearshore habitat have been conducted since 1991; marine bird populations have been surveyed since 1992; and marine bird contamination has been studied since 1995." <http://www.wa.gov/puget_sound/Programs/PSAMP.htm>.

6. RIDGE (Ridge Inter-Disciplinary Global Experiments) 2000: "This plan is the product of three highly interdisciplinary planning meetings attended by more than two hundred scientists. Attendees strongly endorsed the creation of a RIDGE 2000 program that will work towards a comprehensive,

continued

integrated understanding of the relationships among the geological and geophysical processes of planetary renewal at mid-ocean ridges and the seafloor and subseafloor ecosystems that they support. Studies under this new program will be defined by an integrated, whole-system approach encompassing a wide range of disciplines, and a progressive focus within scientifically defined, limited geographic areas." The science plan distinguishes integrated (multiple disciplines focused on one place), exploratory (discovery of new places), and time-critical studies (responding to tectonic events). Each category is addressed in terms of overarching goal (conceptual foundation), questions and hypotheses, and the scope or approach for answering the questions. Technology (measurement devices) and infrastructure (data management) are addressed at the end of the plan. Because the plan was written to motivate federal funding of portions of the plan, there is no explicit description of organizational structure. <http://ridge.oce.orst.edu/R2K/R2Ksciplan/>.

7. "The IPRC (International Pacific Research Center) Science Plan defines the Center's overall structure. It states the IPRC mission, presents four scientific themes and goals, describes specific objectives, and outlines strategies for attaining them." Three of the themes are geographic, focused on Pacific and Indian ocean climate variation, effects of western Pacific Ocean flows on climate, and the Asia-Australian monsoon system. The fourth theme addresses global change as it affects Asia-Pacific climate. The plan includes personnel and infrastructure requirements, and mechanisms for internal management and external guidance. <http://iprc.soest.hawaii.edu/iprc_science/>.

These brief descriptions should make it clear that almost all have at their core a working understanding of the structure and function of a complex environmental system. Surprisingly, many of the plans incorporate long-term change or natural versus anthropogenic change in this conceptual foundation. The following elements are common to many of the plans we examined:

1. The conceptual model and hypotheses to be tested are defined early in the plan.
2. The scope of the plan is defined in terms of place (PSAMP), linkages and flows (SOLAS), or habitat (SALSA, RIDGE).
3. Products relevant to management or plans for outreach are described.
4. Data management strategies are provided.
5. The goals of most programs are expected to be achieved through a combination of long-term research, short-term research, and modeling and synthesis.

These common elements map fairly well onto the elements the committee evaluated for GEM: conceptual foundation, scope, community involve-

> ment, data management, and synthesis and review. We note the lack in most plans of explicit descriptions of organizational structure. This lack probably occurs because the organizational structures are already in place (for instance, in NASA) or because they will never be in place (for many of the science plans that describe loose collaborations). GEM, however, requires an organizational structure to be defined that will disburse funds and involve communities effectively. One other major difference is the size of the plans: Most science plans (with the exception of EOS) tend to be 10-30 pages long. Such conciseness is intentional so that the purpose, scope, and methods can be synthesized down to a clear foundation, and knowing that the scientists involved will work out as the program evolves.
>
> The committee also notes that no plans are designed to involve local communities or traditional ecological knowledge in the formation of research questions and activities. Rather, these plans portray community involvement only through outreach. GEM is in the challenging but exciting position to craft a science plan that bridges science and society in groundbreaking fashion.

Scientific and Technical Advisory Committee should provide scientific oversight and ensure the scientific integrity and quality of the GEM program. An appointed chief scientist or science director should have responsibility for leading and implementing the GEM science program.

The selection of particular projects and observations is achieved through a program's organizational structure, influences who is involved in honing the conceptual foundation into testable hypotheses and research questions, and demonstrates how open the program is to new personnel and ideas. A vibrant and innovative program must encourage new people to become involved over time, yet long-term plans inevitably reward people with previous experience.

Periodic external review of the science program can ensure that the chief scientist and the Scientific and Technical Advisory Committee have the vision and discipline necessary to run a successful program.

4. *Involvement of stakeholders in the planning process.* Large scientific programs designed to understand ecosystems used by a variety of different communities require the support of those communities if the programs are to be of maximum utility. Communities affected by such studies include not just program managers and the scientists involved in conducting research, but also those who live adjacent to the ecosystem, those who harvest resources (whether for subsistence or commercial use) in the ecosystem, and those who use the ecosystem for recreation. When those di-

verse communities can be brought together to plan the studies, rather than just being asked to approve or comment on what others have planned, there is a greater chance of a more holistic view of the goods and services of concern to society and thus the opportunity to design a more satisfactory science program that will enjoy long-term community support.

5. *Management of data to ensure safekeeping and accessibility.* Data management is crucial to a monitoring program because of the need for storing and retrieving large amounts of data (Weisberg et al., 2000). Large long-term scientific studies generate enormous amounts of data, data that must be useful far into the future. One fundamental aspect of data management is that it be designed specifically to support the central purpose of a long-term science program, that is, the comparison of measurements over long periods of time. First, it is essential that there be a mechanism for archiving data that will be durable and that permits data transfer from one storage medium to another as technological innovations appear. A second challenge is to support real-time sharing of data within the program, which is essential for collaboration and integration between disciplines and geographic subdivisions of the study. Third, there needs to be public access to data and data products so the broader community can assess the progress of "their" ecosystem study. Delivery of timely and appropriate data products will be essential if decision makers are to benefit from the program (Weisberg et al., 2000). The successful accomplishment of these three elements makes the data management program the heart of a large long-term scientific program.

6. *Assessment of progress through synthesis and evaluation.* Synthesis and evaluation are essential scientific activities. They provide information on whether a program is making progress toward testing hypotheses and in achieving an understanding of ecosystem function. Syntheses will require a variety of modeling efforts (conceptual, statistical, and numerical), and one should be aware that both the modeling of results and the acquisition of data will vary considerably between physical and biological aspects of the research program (Weisberg et al., 2000). Although generating syntheses of long-term data from these different disciplines is likely to be a challenge, doing so will be important to the long-term success of the GEM program.

This report is divided into sections that address the above elements and includes insights drawn from other long-term science plans regarding issues such as governance structures and data management. Finally, the committee summarizes its conclusions about planning the GEM program and provides recommendations to help guide its continued development.

2

The Importance of a Conceptual Foundation

The stated mission of the Gulf Ecosystem Monitoring (GEM) program is broad and ambitious: "to sustain a healthy and biologically diverse marine ecosystem in the northern Gulf of Alaska and the human use of the marine resources in that ecosystem through greater understanding of how its productivity is influenced by natural changes and human activities" (EVOSTC, 2000a). According to this mission, GEM has a dual purpose—to sustain a healthy ecosystem *and* to ensure sustainable human uses of the marine resources. The second part of the mission statement assumes that these objectives will be accomplished by understanding how both natural changes and human activities influence ecosystem productivity. Implicit in this rationale is that it is possible to separate the causes of natural changes from human-induced changes. It also assumes that a successful monitoring program has to take into account both climate change *and* changing patterns of human exploitation (e.g., fishing practices), which could call for attention to a very complex array of variables.

The GEM program is a long-term monitoring program, and long time series are essential to detecting ecosystem change on intermediate and long time scales. The first step in any research program, particularly one such as GEM, is development of a conceptual foundation, which must be broad because of the program's long time scale. No one can know what theories, taxa, or processes will emerge as critical to the public or managers or relevant to ecosystem functioning in future decades. The choice of a conceptual foundation is critical, as this will drive the choice of species and parameters to monitor. Conceptual foundations that rest on a few indicator species, specific hypotheses about marine ecosystems (e.g., Pacific Decadal Oscillation), or current human impacts (e.g., fishing) are likely to be too narrow and inflexible to support the GEM mission. Instead, the GEM conceptual foundation needs to incorporate the sense that marine ecosystems (processes and taxa) change in response to physical and biological changes and human impacts, as is clearly expressed in the mission statement. Even if the same endpoints for monitoring could be reached by choosing variables to measure in the absence of a broad conceptual foundation (NRC, 1995), it would be difficult to justify them without a conceptual foundation that provides the broad context and helps illustrate relationships.

A solid conceptual foundation will buffer GEM against inevitable shifts in public concerns, such as current concerns with Steller sea lions. Indeed, GEM is aware of the difficulty of pursuing long-term monitoring in the face of short-term interests. There are provisions for multi-decade measurements and for shorter research programs targeting specific issues or hypotheses, so that GEM can respond to current concerns without sacrificing long-term data sets that will prove increasingly useful as they accumulate. A well-designed and broad-based program will provide the best scientific basis for understanding many ecological issues of public concern.

Rendering the conceptual foundation into specific research activities implies the generation of questions. These questions can come from members of the scientific community as well as members of the native communities, fishing communities, state and federal resource managers, and any other stakeholders. The benefits of meaningfully incorporating local communities are twofold: Local knowledge and participation can enrich the scientific program and reciprocally provide a broader basis of support and understanding for the program mission. Indeed, while it is appropriate and probably necessary that a scientific conceptual foundation be developed primarily by scientists, the ability of local communities to inform and provide knowledge of the ecosystem must be emphasized.

Finally, the conceptual foundation must be compatible with the mission of GEM. This mission, as stated in the program, is broad and somewhat indefinite. Despite its breadth, the mission does focus some attention on the reciprocal interactions between humans and the marine environment, although the emphasis is heavily on natural variability, with less attention to measuring human-induced change. Humans derive goods, services, and pleasure from the ocean and consequently, marine systems are affected by these human activities. This occurs in a context of regional climatic and oceanic change—changes that will inevitably and unpredictably occur during the time scale of GEM.

Almost all resource management issues require society to determine the cause of observed system changes. Thus, the conceptual foundation provides a framework for thinking about the kinds of measurements and studies that will be needed if we hope to understand the influences of environmental variation and human activities on the delivery of goods and services from the marine ecosystems. To do this effectively the architects of the GEM program have appropriately taken the long-term view.

The GEM conceptual foundation in the second volume of the August 31, 2000 draft science plan is adequate: It is broad enough to serve over time, is interdisciplinary, and encompasses ecosystem interconnections. It deals with both oceanic and terrestrial ecosystems and the ways that climate and humans influence the production of energy and its flow through these interconnected systems. With a modest restatement, so that it is phrased as an hypothesis, the conceptual foundation could provide a useful guide for research:

> *The Gulf of Alaska, its surrounding watersheds, and human populations are an interconnected set of ecosystems that must be studied and monitored as an integrated whole. Within this interconnected set, at time-scales of years to decades, climate and human impacts are the two most important driving forces in determining the amount of primary production and its transfer to upper trophic-level organisms of concern to humans.*

Given its importance as the guiding force, the GEM conceptual foundation needs to be up front in the GEM science plan instead of in Volume II, Chapter 4. The committee interprets the placement of the conceptual foundation at the end of Volume II as an indication that it is of lesser importance than other elements of the draft science plan. Without a clear and prominent conceptual foundation, it will be exceedingly difficult for the GEM program to remain on course over the coming years as various short-term needs will divert resources and hinder long-term achievements.

The committee is therefore concerned that in the draft science plan it appears that the role of the conceptual foundation in shaping the GEM program has been largely replaced by studies designed to meet short-term needs. There seems to be a critical change in the thinking about the GEM program—from a long-term scientific program driven by a cascade of hypotheses that would determine what, where, and when measurements should be taken—to a program driven by the need to conduct studies in a range of habitats and locations of dubious scientific connection. If this change in emphasis is implemented, GEM is unlikely to fulfill its potential and make unique contributions to improving our understanding of the structure and functioning of a marine ecosystem. We are also concerned that the GEM document gives more emphasis to natural variability as compared to human-induced changes on the Gulf of Alaska ecosystem when both are key parts in the conceptual foundation.

THE SCIENCE PLAN AS A BRIDGE BETWEEN THE CONCEPTUAL FOUNDATION AND A WORKING SCIENCE PROGRAM

A science plan provides the broad outline for translating a conceptual foundation into a working science program by expanding the conceptual foundation into a series of testable hypotheses, questions, or objectives. In the case of the GEM, these hypotheses might concern how energy flows through the various parts of the Gulf of Alaska and Prince William Sound marine ecosystems, and how climate variability at annual to decadal scales might interact with human activities to shape the goods and services obtainable from these ecosystems. Thus, the science plan provides a guideline for the implementation of the GEM program and is the initial guide to scientists, managers, and other stakeholders as they refine the program. While one might not foresee changes in the conceptual foundation of the program, the science plan would be open to modification as new information is gained.

In developing the science plan it may be useful to contrast the ways that we might expect climate and human activities to influence these marine ecosystems. One might expect that climate—through its influences on physical processes as well as through the rates of biological processes through the effects of temperature—will have its primary effects through bottom-up processes that determine the timing, amount, and fate of primary production, including its transfer from one habitat to another. These bottom-up processes are expected to dominate basin and shelf processes, including those in the Alaska Coastal Current. In contrast, one might expect that human activities, through harvest of marine resources including fish, shellfish, and marine mammals, and through the addition of hatchery-raised fishes, will have their primary effects through top-down processes. In the case of the removal of commercially harvested species, the result may be a redirection of energy flow from commercially valuable species (e.g., pollock) to less desired species (e.g., arrowtooth flounder). These impacts are likely to be strongest in inshore and shelf habitats, including Prince William Sound. The other major human impact on this system, pollution, is likely to have its effects restricted to the nearshore, intertidal, and watershed habitats and may exert both top-down and bottom-up impacts. Climate and humans can under some circumstances affect either bottom-up or top-down processes and climate and human impacts may vary in type between habitats. The role of bottom-up and top-down processes in regulating basin, shelf, and watershed ecosystems should be considered when building and implementing a sound GEM science plan.

Questions stemming from the above general hypotheses that might be useful for guiding the development of the core set of measurements could include, for example: How does high (i.e., interannual) and low frequency (i.e., decadal or longer) variation in climate affect the timing, duration, and amount of primary production? How does the timing or duration of primary production influence the fate of organisms dependent on it? What are the fluxes of nutrients and materials between the habitats of interest, and how do these fluxes affect the eventual fate of production in sustaining species of interest to humans? What are the ecosystem-wide effects of the removal or addition of large biomasses of predatory fishes by humans? How does the introduction of pollution affect the ecosystem and how important is the timing, duration, and magnitude of pollutant release? How do fluxes of freshwater, nutrients, and organisms between watersheds and ocean environments affect the dynamics of the ecosystems of the region?

Although there are a number of subsidiary hypotheses presented in Chapter 4 of the GEM document (EVOSTC, 2001), there is little effort to tie them into the program's conceptual foundation or to explore how they might provide the connections needed between the conceptual foundation and the development of the science program. Thus, the GEM team has not used the conceptual foundation to develop its research plan. The conceptual foundation provides a clear, concise framework of the functioning of the Gulf of Alaska and Prince William Sound marine ecosystems. If the GEM plan is to be coherent and successful over the long term, the conceptual framework must be at the center of the program, with all research and monitoring emerging from and addressing it.

The development of the science plan from the conceptual framework will benefit from a review of existing data. Such a review should take advantage of the many years of research funded by the *Exxon Valdez* Oil Spill Trustee Council, as well as the results of the many independently funded research activities that have occurred in the northern Gulf of Alaska and adjacent waters. These syntheses should include investigation of what has been learned about ecosystem function in the Bering Sea, other areas of the North Pacific, and in the sub-Arctic seas of the North Atlantic Ocean and the Barents Sea. The hypotheses used to focus GEM's long-term research will set the course of the program for many years to come. Deciding on the best approach is not something that should be done quickly or without benefit of other programs. A carefully crafted conceptual framework and attendant hypotheses will determine the success or failure of the program.

A broad conceptual foundation with a sound scientific basis provides a strong scientific justification for the program. It provides an intellectual structure that can guide modification of the program if that becomes necessary. One might ask if this approach is too academic for a program that includes applied management goals and whether it would preclude the study of issues identified by managers or the public. The opposite is true. If the GEM program has a broad scientific foundation, then short-term issues of public concern can be addressed as elements in this broad construct. Even more important, a sound scientific framework would make it much more likely that the GEM program will collect the most useful and important ecological information. However urgent an environmental issue might be, understanding and managing it almost always depends on scientific understanding. Thus, a soundly designed program based on a scientific conceptual foundation should not be seen as an alternative to local community and public concerns. Instead, it should be recognized as the only way to do that effectively over the long term. The committee offers the following recommendations to achieve this broad goal:

- The science plan should include a broad conceptual foundation that is ecosystem-based. It should seek to understand natural and human-induced changes and it should be flexible to accommodate changing needs without compromising core long-term measurements.
- The GEM science plan should articulate two or three fundamental hypotheses about the ecosystem that then should be used to guide the selection for monitoring of particular species and other physical, biological, and human aspects of the ecosystem.

3

Determining Scope and Geographic Focus

SCOPE

Three interrelated elements must be defined when setting the scope of a science plan in order to focus attention and resources on a practical subset of the vast array of possible research questions. The first two elements, geographic focus and research approach, serve to set bounds on "where" the plan is applied. The geographic focus delimits the spatial extent of the plan. The research approach is the decision about how to divide research efforts in the geographic area. For instance, based on the program's main goals planners might elect to give disproportionate attention to particular habitat types, species, flows of energy or materials, or the consequences of specific perturbations. The third component of scope is determining generally "what" will be measured, which follows once the first two elements are agreed on and involves the selection of core long-term variables to measure.

GEOGRAPHIC FOCUS

When resources are finite, there are inevitable tradeoffs between the intensity and geographic focus of research. Multiple variables can be monitored in a small area, but only a few are feasible to monitor at multiple locations. The choice of geographic scale for a long-term science plan should include the following considerations:

Scientific criteria. Is the scale relevant to the hypotheses of interest? Specific questions about human-induced and other changes can be framed at a variety of scales. For example, at relatively small scales: How does the consumption of intertidal herbivores by humans affect algal production? At relatively large scales: Is offshore production, as indicated by chlorophyll, related to the nesting success of seabirds? According to its title, the GEM plan takes the Gulf of Alaska as its scope. However, the central hypothesis of the plan—that natural and anthropogenic factors interact to influence biological productivity—could be addressed at a variety of scales in the Gulf of Alaska.

Building on the knowledge base. As a new research program is developed it can build on past work in three ways: (1) by continuing past work (extending the time frame), (2) by collecting information on unstudied variables (extending the intensity), or (3) by collecting information in unstudied locations (extending the spatial scale). The choice among these options requires that existing data be synthesized first. Many of the natural changes in the Gulf of Alaska are thought to cycle at intervals of several decades. Because little monitoring has been ongoing for such long periods, continuing past measurements may represent the most effective way of testing for variation at this temporal scale. Second, if two existing measurements show striking correlations, measuring new variables can be an effective way of testing the mechanisms of interaction among complex environmental factors. For instance, if ocean survival of salmon varies with phytoplankton production, then measuring forage fish abundance and demography could provide an intermediate food-web linkage. Finally, extending the spatial scale of measurements is important for determining the generality of hypotheses that have previously been tested only locally. This last choice in particular requires adequate synthesis of existing data; otherwise, it is impossible to ask whether existing patterns are general (because there are no existing patterns).

Management needs. Although GEM's mandate is not resource management, most large science programs are justified in part by the usefulness of products provided for decision makers (Weisberg et al., 2000). Most management issues are fundamentally local because this is the scale of human impacts (barring atmospheric change); however, the precise locations where prior data would be useful can shift over time. For instance, baseline data in Prince William Sound would be useful if another oil spill occurred there but it would not address eutrophication in Cook Inlet. A broad geographic scope can improve the chances that long-term measurements remain relevant as management issues change.

Accessibility and cost. Cost is the basic limitation setting the tradeoff between intensity and scale of monitoring. One drawback of a large geographic scope is that tremendous resources are required simply to travel to research sites. Travel costs may be reduced if monitoring is carried out in local communities and if automated data collection is used for basic measurements. Many hypotheses can be tested using a variety of methodologies, variables, or research sites. For instance, Pajak (2000) proposed 13 fundamental ways to measure ecosystem sustainability, incorporating ecological and social considerations, and provided six variables that would be suitable for each. It follows that cost could be used as a criterion for choosing among monitoring sites or variables with similar ecological importance.

The GEM plan has taken the northern Gulf of Alaska as its geographic scope. In its interim report the committee recommended that GEM initiate long-term research in Prince William Sound, then extend geographic coverage over time. The rationale underlying this recommendation was the difficulty of designing a useful research plan for a broader area given limited funds, coupled with the utility of extending time series at the core of the area affected by the spill in 1989. The Trustee Council is well within its prerogative to select any geographic scope, however, if the program is to be successful, the scope should be justified on science and management grounds and must be appropriate to the funding level.

Although it is possible to justify a focus on the entire Gulf of Alaska given the above criteria for selecting geographic scope, the committee is concerned that the geographic scope has been chosen primarily to be sure that all stakeholders get a "piece of the pie." Covering a large geographic scope in the absence of a scientific rationale (unifying framework) risks dividing resources in a piecemeal fashion that will make synthesis and interpretation difficult. Indeed, this problem is epitomized by the list of interim projects in GEM planning documents. There is a strong geographic focus on Kachemak Bay and Cook Inlet, for instance, which may reflect the distribution of humans along the coast rather than addressing core hypotheses. In addition, existing oceanographic measurements (GAK1 hydrographic station, ADCP current measurements at Hinchinbrook Entrance, thermosalinograph and fluorometer on a tanker, and thermosalinograph on a Kachemak Bay boat) are not obviously linked to the three projects on modeling ocean circulation.

A politically motivated scope is particularly detrimental to long-term monitoring if the projects focus intensely on particular areas for short periods of time. If GEM activities are directed by current management concerns, it is likely that the geographic focus will be buffeted, and the monitoring will fail to provide the long time series it is uniquely poised to generate. If the geographic scope remains as the entire Gulf of Alaska, it is imperative that the choice of variables to measure be made with extreme care.

The Gulf of Alaska is an area of about 1.2 million km^2 and the continental shelf in the Gulf of Alaska is 0.37 million km^2, about 10 percent of the entire U.S. continental shelf area (Hood, 1986). GEM is projected to provide about $6 million annually for research and staff to facilitate science and education (<www.oilspill.state.ak.us/future/future.htm>). Although this is a sizeable budget, the area to be covered is quite large. Other large programs in marine science provide an instructive comparison (Table 3-1). The focus of each of these programs is much more targeted than is GEM, yet most have more money to spend on a per-area basis (Table 3-1).

HABITATS AS A DIVISIONAL UNIT

Because of the tradeoff between geographic scope and intensity of research effort, science plans covering large areas must include methods for stratifying observations and allocating funds for short-term process studies. This focus can be provided in a number of ways.

1. *Flows of energy, impact, or materials*. The plan could focus on one or a few important flows through the geographic area, for instance, across-shelf transport or movement of pollutants through food webs.

2. *Habitats or regions*. The plan could foster research in smaller areas that are believed to be representative of a broader region or habitat type.

3. *Species*. The plan could focus on one or a few species throughout the geographic area.

4. *Hypotheses*. The plan could target research toward a restricted hypothesis, for instance, taking measurements that would support or disprove the Pacific Decadal Oscillation as a cyclic climatic shift.

5. *Time*. The plan could incorporate intentions to develop research projects in different areas over time. This strategy would approximate that of the U.S. Environmental Protection Agency's National Estuary Program (<www.epa.gov/nep>), which provides funds to develop management plans in one estuary after another. This strategy is generally inappropriate when the plan's mandate is to generate consistent long-term data sets.

Of these options for stratifying observations, habitat is perhaps the most widely used approach. Division by habitat has one clear advantage for GEM implementation: It clarifies the amount of money being spent close to and far from shore. The GEM plan articulates a rationale for focusing on nearshore observations and studies: This area is relatively unstudied, and people living along the coast interact with it directly.

Division by habitat has several problems. In the GEM document, hypotheses are presented as repetitive questions listed for each habitat type, but they would need considerable refinement before they could be a useful guide for research. For example, the GEM document asks the same

TABLE 3-1 Comparison of Funding Levels for Large Marine Research Programs

Program	Annual Funding ($)	Shoreline Length (km)[a]	Annual Funding ($ per km)	Area (km^2)	Annual Funding Per Area ($)
GEM[b]	6×10^6	1,500	4,000	1.2×10^6	5
PISCO[c]	5.75×10^6	2,000	2,875		
GLOBEC[d]	3×10^6	250	12,000	48,000	62
SEA[e]	3×10^6			38,000	80
Chesapeake Bay[f]	12×10^6	7,000	1,700	5,900	2,000

[a] For these different programs, the method for determining shoreline length is inconsistent, so these comparisons are approximate. GEM and GLOBEC are done similarly but the others might be determined using fractals that can make the length a less dependable number.

[b] GEM shoreline length measured on map; annual funding estimated.

[c] PISCO (Partnership for Interdisciplinary Studies of Coastal Oceans) addresses benthic-pelagic coupling on rocky shores in California and Oregon. Shoreline length from <www.piscoweb.org>; annual funding estimated.

[d] GLOBEC (Global Ocean Ecosystem Dynamics) focused on a small area of the Gulf of Alaska. Shoreline length measured on map; annual funding estimated.

[e] SEA (Sound Ecosystem Assessment) was a major portion of EVOSTC-funded research, developed in 1993 and ran for seven years. Information from GEM program and <www.oilspill.ak.us/research/resrch.htm#SEA3>.

[f] Chesapeake Bay shoreline length from <222.gmu.edu/bios/bay/cbpo/into.htm>; funding level estimated by committee.

questions for continental shelf and nearshore areas, although these areas have different natural and anthropogenic forcing functions (Table 3-2). Most importantly, the habitat divisions may set up a barrier to understanding links and transfers among habitats. The committee cautions against the development of habitat-based subcommittees in the organizational structure, as there is substantial risk of neglecting linkages among habitats in setting research goals.

Table 3-2 reproduces, in tabular form, the habitat-specific questions that form the core of the GEM plan (vol. 1, ch. 3). These questions actually begin to develop a set of hypotheses about how natural and anthropogenic factors influence ecosystem functioning, recognizing that different factors may be important in different habitats. As these hypotheses are refined by a scientific steering committee, they could help guide the selection of long-term observations and process-oriented research.

The committee discussed these working hypotheses in some detail, and it offers a few observations about the current framework. These ob-

TABLE 3-2 Current Hypotheses About Natural and Anthropogenic Forcing Functions in Four Gulf of Alaska Habitats as Provided in Volume 1, Chapter 3, of the GEM Plan (EVOSTC, 2001)

Habitat Type	Natural Forcing Functions	Anthropogenic Forcing Functions	Habitat Variable of Interest
Watershed	Climate	Habitat degradation Fishing	Marine-related production (nutrients from salmon)
Intertidal/subtidal	Currents Predation	Development Urbanization	Community structure and dynamics
Alaska Coastal Current	Strength, structure, and dynamics of the Alaska Coastal Current	Fishing Pollution	Production of phytoplankton, zooplankton, birds, fish, mammals
Offshore	Alaskan Current/ Alaskan stream Mixed layer depth Wind stress Downwelling	Pollution	Carbon production and shoreward transport

servations are not meant to be prescriptive; they simply point out areas that require additional consideration. Some of the forcing functions are not parallel. For instance, "climate" is hypothesized to affect watershed production, but more specifically "wind stress, mixed layer depth, and downwelling" are hypothesized to affect production offshore. Some of the habitat variables of interest, which should reflect ecosystem functioning, are too general or inclusive to measure. Specifically, "production of phytoplankton, zooplankton, birds, fish, and mammals" would require monitoring all taxa in the coastal region.

Similarly, "community structure and dynamics" in the intertidal/subtidal zone provides no indication of which taxonomic groups are expected to be most sensitive to change or most important to human communities. The metrics most sensitive to perturbations or stresses may not be abundance but the size or age structure of populations (Paine et al., 1996; Driskell et al., 2001; Monson et al., 2000).

The Alaska Coastal Current travels through a relatively narrow band (< 50 km) of the coastal region of the Gulf of Alaska, so it would be useful to use two different habitats instead: (1) the nearshore to 50 km, including bays, sounds, and the Alaska Coastal Current; and (2) the continental shelf that extends from the nearshore to the shelf break. Finally, it is possible to incorporate across-habitat linkages by developing hypotheses about how different habitats may be strongly coupled or the degree to which they behave independently.

Table 3-3 provides a refined set of hypotheses about how natural and anthropogenic forcing functions and across-habitat linkages may influence biological production. We emphasize again that this framework is not prescriptive but is provided to illustrate how study of linkages might be accomplished. These kinds of refinements should be made as the plan develops, using existing scientific data to justify choices of most important forcing functions. Both the forcing functions and "habitat" response need to be measured to test the underlying hypotheses.

CHOICE OF VARIABLES AND RESEARCH PROJECTS

The three types of research included in the GEM plan—measuring variables over the long term, carrying out shorter-term studies of processes, and synthesizing and analyzing collected data sets—will require different strategies for implementation (from the call for proposals to the selection process to the evaluation phase). Recognizing that many large scientific programs focus on just one or two of these types of research, it is clear that GEM planners will face challenges giving appropriate weight to each type and designing implementation strategies for each. Important points for GEM planners to consider for each type include:

- Long-term research requires a large amount of up-front effort to choose variables. Determining who carries out long-term research is particularly difficult because it cannot (and should not) be assumed that the same research group will collect the information for the next 100 years. Data collection efforts should be evaluated on the order of every five years. Sampling protocols should be kept as constant as possible and if changes in technology occur, ample attention should be paid to inter-calibration of the time series.
- Short-term process studies will give the GEM program some of the flexibility it needs; typically, requests for proposals for this type of work occur every one to two years, so that the focus can be changed in accordance with steering committee and community interests.
- Synthesis should be an ongoing effort, some of which will involve modeling. Invitations for proposals should occur every two to four years, and a postdoctoral program might be an excellent way to have long-term data sets analyzed in novel ways (for instance, see the National Center for Ecological Analysis and Synthesis postdoc program at <http://www.nceas.ucsb.edu/frames.html>).

TABLE 3-3 Potential Habitat Divisions in the Gulf of Alaska and Hypotheses About Most Important Factors Influencing Biological Production

Habitat Type	Natural Forcing Functions	Anthropogenic Forcing Functions	Strongest Across-Habitat Links	Habitat Variable of Interest
Watershed	Rainfall Offshore production	Habitat degradation Fishing	Salmon returns	Marine-related production within watersheds
Intertidal/subtidal	Predation	Shoreline development Pollution Direct exploitation	Larval and food delivery from continental shelf	Recruitment and species interaction strengths
Nearshore, including Alaska Coastal Current	Wind stress Freshwater	Fishing Pollution	Freshwater input	Biomass and production of phytoplankton, zooplankton, and forage fish
Continental shelf	Resupply of nutrients Currents Mixed layer depth	Anthropogenic climate change	Across-shelf flows	
Offshore	Mixed layer depth Wind stress	Anthropogenic climate change	Across-shelf flows	Phytoplankton production and shoreward transport

Balancing Long- and Short-Term Research

Long- and short-term studies differ in their focus and their funding requirements. A research plan that aims to fund both, as the GEM program does, must decide how to balance resource allocation to best meet its program goals. The present GEM draft plan does not address this critical issue. The term "monitoring" has always been in the title of the GEM plan, and the committee believes this focus on long-term monitoring should remain central to the GEM program. Many of the biological and physical processes of interest to GEM operate at decadal or longer temporal scales, and require long-term measurement if patterns and variability are to be evaluated.

The ability of GEM to support long-term marine ecosystem studies is essentially unprecedented. No other current programs have this capability, nor are they likely to. In contrast, there are numerous funding sources for short-term research projects. The committee recognizes that short-term studies can be valuable for optimizing long-term study design. For example, they might be used to evaluate which of several techniques are most appropriate for remote sensing of nearshore measurements. The committee feels the GEM program should start out by devoting the majority of its resources, perhaps even all of them, to setting up and maintaining the long-term research program, with few resources used initially for short-term research. (Resource allocation is discussed in more detail in Chapter 4.)

Strategies for Effective Choice of Long-Term Measurements

A well-crafted, long-term research plan addresses the program objectives as defined in a mission statement and a conceptual foundation. Although spatial and temporal scope (i.e., where to conduct measurements and for how long) may be settled in many ways, the core variables (what to measure and how often) usually flow from hypotheses and models. A comprehensive database of existing research results can aid in the development of these hypotheses. For effective management of coastal resources, monitoring programs must collect data at multiple scales, and most importantly, must link measurements between these scales, an often difficult process (Weisberg et al., 2000). Such linkages are necessary to provide managers with predictive models of the interrelated processes underlying ecosystem function to support wise decisions for managing resources.

Because of the long time frame of GEM, it is critical that the core variables for monitoring be chosen with great care. The GEM plan outlines a general strategy for identifying these variables and implementing the monitoring program (Figures 3-1 and 3-2). This strategy shows that GEM's mission and goals imply a broad conceptual foundation, from which will emerge hypotheses. Research to address these hypotheses will be carried out if similar work is not already being done. In short, hypotheses and questions get priority, and the plan recognizes the utility of asking whether existing data can address these questions before embarking on entirely new data collection. The committee agrees with this general strategy.

The role of synthesis. The GEM plan is inconsistent in exactly how synthesis fits into the choice of long-term variables. Selection of long-term measurements may include some modeling (EVOSTC, 2001, vol. I, p. 37 - "Initial synthesis activities, including modeling, would support identification and development of testable hypotheses."). Data synthesis is identified as preceding research in some parts of the text (EVOSTC, 2001, vol. I, p. 37 – "Synthesis—Research—Monitoring"), but is listed as concurrent with research in other sections (research and synthesis are identified as concurrent activities in 2003, the first year of plan implementation). What is an appropriate order?

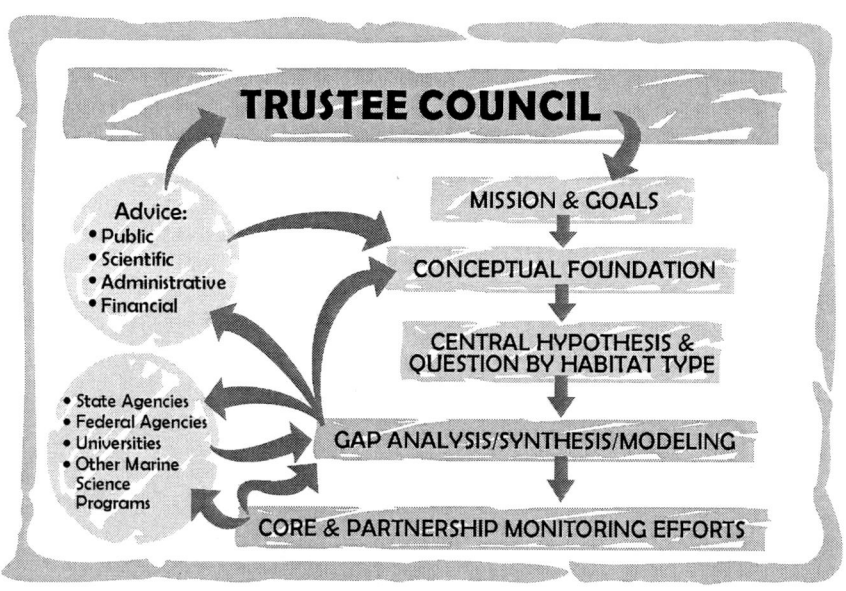

FIGURE 3-1 In the GEM plan selection of the variables to be measured starts with the mission and goals established by the Trustee Council, as expressed in the conceptual foundation, and is developed with input from numerous sources (EVOSTC, 2001, vol. I, p. 38).

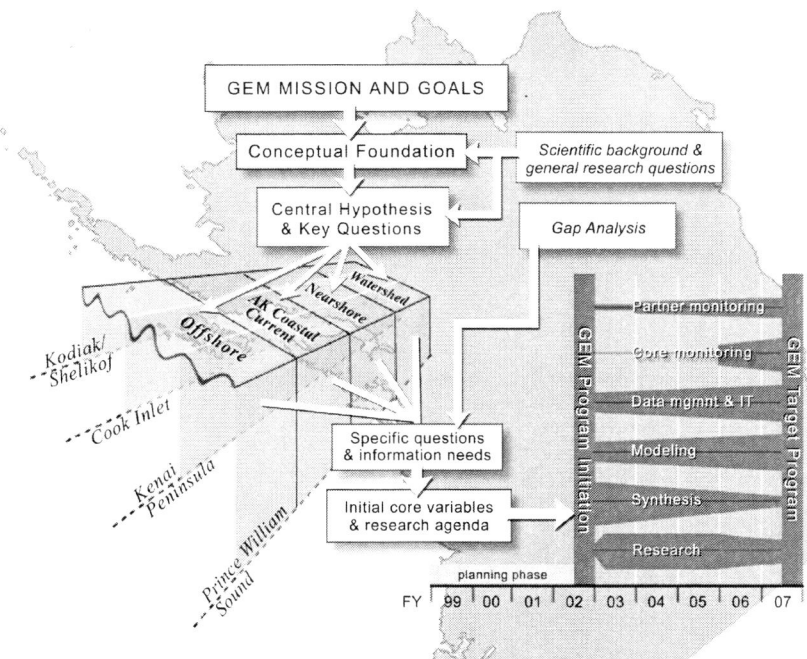

FIGURE 3-2 A schematic overview of the structure of the GEM draft science plan, showing the relation of key concepts to the habitat and the schedule of implementation (EVOSTC, 2001, vol. I, p. iii).

1. Hypotheses can precede synthesis; indeed, they can help guide it.
2. Some variables for long-term measurements may need to be chosen before synthesis is complete, because synthesis should continue through the life of GEM.
3. Data synthesis must be included in an ongoing process throughout the life of the GEM program to optimize identification of additional variables for both short- and long-term projects.

For the GEM program enormous amounts of data already exist on the physical and biological features of the Gulf of Alaska, much of which has been generated by Trustee Council-supported research undertaken since the *Exxon Valdez* oil spill. At present these data have been gathered but have not been synthesized into a comprehensive, easily accessible database. Creation of such a database should begin immediately, with rapid updating of data in a readily usable form. (Approaches to data synthesis and model building are discussed in more detail in Chapter 7.)

> **BOX 3-1**
> **Markers of Ecosystem Health**
>
> Parameters or markers associated with ecosystem health have been used in numerous monitoring programs such as the Bermuda Atlantic Time Series (BATS), Hawaii Ocean Time Series (HOTS), and California Cooperative Fisheries Investigations (CALCOFI). GEM should look to these programs for guidance in choosing such markers, keeping in mind that some indicators may not be appropriate for the Gulf of Alaska ecosystem. For example, biodiversity has been used as an indicator of ecosystem health in many programs but may not be appropriate for high stress environments. In Alaska rapid colonizers may be wiped out catastrophically by winter storms, yet return the following year. Such natural patterns in community structure must be distinguished from anthropogenic effects for biodiversity to be a useful indicator of ecosystem health in the Gulf of Alaska.

The role of workshops. Identification of suitable variables for long-term research will in the end be carried out by the steering committee as it develops proposal solicitations and evaluation criteria. While these proposal invitations must be derived from GEM's conceptual foundation to maintain program focus, it is critical that community input be incorporated into the proposal solicitation at this early stage of the program. Two ways that substantive community input could be obtained would be through the Public Advisory Committee and by holding a series of workshops covering variables for long-term measurements. Workshops are not included in the plan but do appear to be funded this year (e.g., concerning herring, ocean circulation, and intertidal monitoring as described in EVOSTC [2001], vol. I, p. 56). It is unclear whether they will include community, manager, and researcher participation.

Valuable metrics of long-term change are those most sensitive to climate and/or anthropogenic trends or perturbations. In this regard GEM might also consider variables that serve as markers of ecosystem health. Such markers have been used in other long-term research programs (Box 3-1).

Implementation of the Plan

Proposal solicitations based on the conceptual foundation and designed by an integrated group of scientists and community stakeholders will ensure that both quality science and issues of relevance to the community are incorporated into the plan. Selection of those proposals that best address the solicitation will ensure that the variables most sensitive to changes in the system, and most relevant to the program's goals, are chosen for long-term measurement. Data synthesis must be seen as an ongoing process and provisions made to ensure timely incorporation of new data into the database. A commitment to timely data synthesis will facilitate timely recognition of patterns and their normal range of variability. If long-term baseline data had been available for more species in the Gulf of Alaska at the time of the spill, managers would have been able to determine whether shifts in population densities were due to the spill and cleanup efforts or simply reflected population trends already in progress at the time of the accident.

Concerns About Choice of Variables

The choice of variables to monitor should not be done exclusively through gap analysis or by partnering with existing programs. Selection procedures need to address how often and where variables will be measured at the same time that particular variables are chosen. Effective implementation of the strategy for selecting variables, which we believe needs to address community interests, will be difficult. Elaboration of these concerns follows.

Partnering. The success of any long-term research program ultimately depends on an unwavering commitment to repeated measurement of a set of core variables that is not altered over the life of the program. While variables may be added, core variables must never be dropped or the usefulness of the long-term data set will be compromised. In this regard, GEM should not rely on partnering with other scientific programs for collection of any core variables. These programs will invariably be shorter-lived than GEM, and have different goals and foci.

Gap analysis. The GEM Draft Plan proposes the identification, and filling, of gaps in our knowledge base (gap analysis) as a critical step for identifying core variables (Figure 3-2). While the committee acknowledges the need for basing decisions on a comprehensive, scientific database of the Gulf of Alaska, filling gaps without hypothesizing how the resulting data specifically relate to the conceptual foundation runs the real risk of expending resources to generate data of little relevance to the program. There will always be information gaps, and as we learn more about the system, more gaps will be identified. Whether or not filling these gaps is necessary can only be determined using a hypothesis-based approach.

An example of what may happen using the gap analysis strategy as outlined in the GEM Draft Plan is that measurements of temperature and salinity might be identified as high priority. Regions within Prince Will-

BOX 3-2
The Evolution of Major Science Plans Takes Time

The creation of all long-term science plans takes time because the process of developing the plan is as important as the details included in the plan. For example, the U.S. portion of the Joint Global Ocean Flux Study (JGOFS) had its beginnings in 1984, with the international component starting about three years later (NRC, 1999b). The formation of this effort was not simple.

Initially, the U.S. Global Ocean Flux Study (GOFS) was an outgrowth of three separate science community projects that were active in the early 1980s: The National Academies' Ocean Studies Board was investigating the feasibility of a program that would conduct long-term studies of the biological and chemical dynamics of the ocean on basin-wide and global scales; the NSF Advisory Committee for the Ocean Science Program was developing a long-range plan; and a separate National Academies committee had identified initial priorities for the International Geosphere-Biosphere Programme. As the relationships among these activities became clear, and with support from NSF, NASA, ONR, and NOAA, a group of scientists met in 1984 at Woods Hole under the auspices of the National Academies. This generated the basic scientific underpinnings that defined the proposed mission for GOFS and led to the GOFS Scientific Steering Committee, which was formed in 1985. Then, after continued discussion and planning, in 1987 an overview document was published that more fully outlined the program. Between 1986 and 1990, the science community produced nine reports that summarized the recommendations of workshops designed to expand on the general plans, covering topics such as water column processes, benthic processes, continental margins, data management, and modeling. Finally, in 1990 the JGOFS Long Range Science Plan was published, based in part on the recommendations of the workshops. It was 1995 when JGOFS released an Implementation Plan, which gave the status of the JGOFS research and future directions.

One strength of a major research program is the ability to draw and direct a significant amount of talent and scientific interest toward a large and often high profile scientific challenge. But to realize that opportunity requires significant advance planning and coordination, and one key element is taking the time necessary to allow wide participation in the program's definition and evolution.

SOURCE: NRC, 1999b.

iam Sound such as College Fjord might be identified as locations where no such measurements have been done. Thus, lack of temperature and salinity data in this area would be identified as a knowledge gap and given high priority. If the location was populated with people and marine mammals, this area might become the highest priority for gap analysis. These measurements might be prioritized because they would be less expensive to collect relative to similar measurements taken in a remote region offshore on the continental shelf. However, such sampling within the fjord would not necessarily lead to a better general understanding of marine processes.

Community involvement. Communities can play a significant role in generating scientific ideas that are relevant to the goals of the GEM program. The culture and livelihood of local stakeholders often depends on the health of the ecosystem. Their intimate knowledge of the dynamics of the system, based on daily, and often generational, experience (e.g., changes in predator and/or prey abundance in response to climate change or to the introduction of hatchery-reared fish) can significantly broaden the range of research questions and approaches. Incorporation of meaningful community involvement in the *generation* of scientific questions for a research plan of GEM's scope and duration would significantly enhance both the quality of the science and its relevance to the community. Further, involved citizens whose efforts and contributions are meaningfully incorporated into the plan are more likely to provide strong support for the program for the future. Finally, the concerns of stakeholders often reflect the concerns of managers. While many of these concerns can best be addressed by the long-term research program, some may reflect specific issues or hypotheses that require more immediate answers. These could be addressed by incorporating short-term studies (3-5 years) into the monitoring program, thereby allowing GEM to respond to current concerns without sacrificing long-term data sets that will prove increasingly useful as they accumulate. A research plan that incorporates meaningful community involvement would serve as a model for other programs grappling with how to address the concerns of resource managers and local communities into their science plans. (The value of community involvement is further discussed in Chapter 5.)

Implementation. Finally, how the program will be implemented must be made clear. The roles and responsibilities of each participant and committee must be clearly defined, and the paths of information flow outlined, to demonstrate how the program will operate in practice. The design of long-term programs can take several years (Box 3-2), however, a carefully designed plan is well worth such an investment. Collection of the wrong data, poor program management, or other flaws in the plan could seriously jeopardize GEM's credibility and erode long-term support for the program.

4

Organizational Structure

Major marine ecosystem programs require a large commitment of human and fiscal resources, and the assurance of scientific credibility and coordination are essential. The effectiveness and character of marine ecosystem research and monitoring programs are greatly influenced by their organizational structure, because it is the structure that ensures that the goals of the science plan are translated into specific research activities. A credible scientific program must be structured so that program planning and review, implementation, community involvement, coordination, proposal solicitation, peer review and funding, interactions among investigators, data management, oversight, and public outreach all are facilitated efficiently.

Most interdisciplinary marine ecosystem programs have a scientific steering committee and a chief scientist (or scientific director) that together develop and implement the science plan and provide program oversight (Figure 4-1). In this science management structure, the chief scientist (who serves as an ex-officio member of the steering committee) works jointly with the steering committee and is empowered to develop and implement the program science plan. The chief scientist has authority regarding all scientific decisions after consultation with the program principal investigators and the steering committee. The chief scientist must concentrate on developing and implementing the program science and informing the interested communities of program results. To allow time for these scientific activities, the program's scientific administrative duties are usually delegated by the chief scientist. The chief scientist of interdisciplinary science programs similar to the Gulf Ecosystem Monitoring

(GEM) program are normally scientifically well-rounded investigators who are respected nationally and internationally by their peers. The *Exxon Valdez* Oil Spill Trustee Council should seriously consider the adoption of a similar organizational scheme. The recruitment of suitable candidates might be made easier if there were a relationship of the individual with a university.

The GEM program implementation plan envisions that interactions between the Public Advisory Committee, Scientific and Technical Advisory Committee, and the general public, along with an external GEM program review every five to seven years, will provide the needed scientific oversight. The committee agrees that the chief scientist working with the Scientific and Technical Advisory Committee (which is, in essence, the "steering committee" referred to above) and the Public Advisory Committee should play a key role in program oversight. If GEM is to succeed, its oversight activities must address issues such as the preparation of science and program implementation plans, proposal solicitation and peer review, investigator information exchange, program data management and outreach to Alaska natives and other communities of interest. The Scientific and Technical Advisory Committee, working with the chief scientist, should play the dominant role in assuring GEM scientific program credibility and direction.

Science planning must continue during the life of the GEM program to assure program success. Initially the core variables to be monitored must be carefully selected and should not be modified without careful consideration during the life of GEM. This will assure that consistent long-term data are obtained with a principal objective of distinguishing between human-induced and natural changes in the Gulf of Alaska ecosystem. A monitoring subcommittee reporting to the Scientific and Technical Advisory Committee may be of value in both developing monitoring protocols and requests for proposals, but such a committee should not be the sole mechanism by which the variables to be monitored are selected. The GEM program as a whole should be involved with the selection of variables to be monitored. This might be achieved through a series of targeted workshops to assist the chief scientist and/or Scientific and Technical Advisory Committee in determining location and frequency of measurements needed to monitor key biological, chemical, and physical variables. The importance of the early synthesis to the long-term success of GEM cannot be overstated.

The GEM program must develop a clear implementation plan that includes some well-defined milestones and coordination among the agencies and programs conducting short- and long-term ecosystem research in the Gulf of Alaska. The plan should provide for an iterative assessment

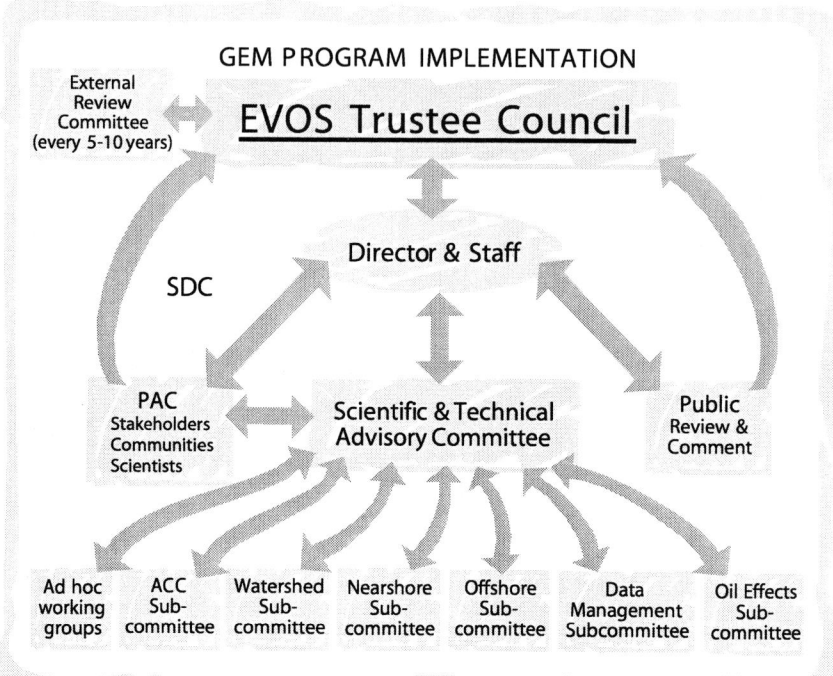

FIGURE 4-1 This figure describes the proposed decisionmaking and management structure for implementing the GEM program document and the GEM monitoring and research plan. Information and guidance flows between the Trustee Council and the Program Advisory Committee, the Scientific and Technical Advisory Committee, and the public at large, through the executive director and staff. The six-member Trustee Council makes all funding, programmatic, and policy decisions. All decisions must be unanimous. The Trustee Council relies on the executive director and staff to ensure that decisions are implemented and that the advice and review from the Program Advisory Committee, the Scientific and Technical Advisory Committee, and the public are organized and summarized to assist in its decision making. The Program Advisory Committee, which is required by the settlement to be established under the Federal Advisory Committee Act, consists of stakeholders, scientists, and community representatives who meet at least twice a year to provide advice and feedback to the Trustee Council on the overall direction of the program, including proposals to be funded. The Program Advisory Committee takes an active role in setting priorities and ensuring that the overall program is responsive to public interests and needs. The Program Advisory Committee is not intended to be the only conduit for public input. Additional public advice is sought on a regular and formal basis from the public at large, including public notice of all meetings, regular opportunities for public comment, and public hearings. The Scientific and Technical Advisory Committee provides key technical review and advice for the program, both from the "bottom

and evaluation of program objectives. Program reviews, both internal and external, should include:

1. evaluation of progress made toward the scientific objectives;
2. recommendations for any needed changes to scientific goals and the implementation plan;
3. identification of opportunities for greater involvement of scientific, native, and local communities in planning and implementing of the GEM program; and
4. reporting of GEM results to relevant scientific and Gulf of Alaska communities and GEM sponsors.

The GEM organizational structure must include procedures for efficiently soliciting and evaluating research proposals. Not only the scientific community but also other communities, such as Alaska natives and commercial fishers, need to be a part of the GEM management of proposal solicitations and funding approval. These communities require an effective way of submitting quality proposals addressing their needs. GEM should actively recruit participation of these communities to assure program openness and that its foundation is built on the broadest community base. Proposal reviews should have a peer review foundation. GEM staff and GEM-funded scientists may serve as proposal reviewers, but additional peer reviewers not employed or funded by GEM should evaluate each proposal. The GEM program will require solicitation of proposals to collect specific required core measurements along with those solicited to conduct innovative science. GEM must assure that the core measurements are collected efficiently and consistently on an ongoing basis. Sufficient resources should be available for sample processing (e.g., species identification and enumeration) in a reasonable period of time. The funding of the core measurements must receive the highest priority and may require the majority of GEM funds.

---

up," using a group of subcommittees organized by habitat and other functions (e.g., data management), and from the "top down," by a core committee composed of subcommittee chairs and other distinguished scientists and technical experts. The subcommittees help develop testable hypotheses, identify core variables and monitoring stations, and assist with peer review of proposals. The core committee ensures that the program is comprehensive across all habitats in working to answer the central questions and hypotheses. In addition, the Trustee Council is advised by an independent External Review Committee convened at the request of the Trustee Council, which at least once every five years conducts a review of the GEM program.

The GEM organizational structure will need to direct over time the issue of the balance between long-term monitoring and process studies in the GEM program and the associated funds devoted to each of these activities, as the allocation of funds is not explicitly discussed in the GEM strategic plan. Given the funds that will be available over the first decade, it is unlikely that the long-term monitoring program could be achieved unless a major fraction of funds is committed to this activity. It is very likely that the desired monitoring program could require the entire budget, because monitoring costs include data collection, data processing, and electronic data storage, and maintenance. The costs of data processing, storage, and maintenance should not be underestimated or undervalued. The longer-term success of the program will depend heavily on the early and continued commitment to all components of monitoring.

This means that the decision to fund short-term process studies will need to consider the extent to which such studies may jeopardize long-term measurements. GEM managers should expect that establishing and implementing the long-term monitoring plan will dominate the early years of the GEM program and that process studies will play a larger role once the long-term measurements are in place. Over the longer term the balance between long-term monitoring and process studies should be guided by the GEM goals to detect and understand changes in marine ecosystem structure and functioning, as a basis to inform, solve, and predict the consequences of these changes. To be true to its mission and to achieve GEM goals, the monitoring component cannot be compromised and must be the GEM program centerpiece.

The GEM organizational structure must make certain that data management receives serious and consistent attention. The importance of data management and data archiving cannot be overemphasized given the long-term objectives of GEM (see Chapter 6). Program leadership must track data management progress effectively; and a comprehensive data management group is the best way to accomplish this. An effective data management subcommittee could play a key role in assuring that data management and archiving are effective and efficient. Proper data management will make data easily available for analysis, synthesis, and modeling exercises conducted throughout the life of the GEM program.

The GEM organizational structure must include mechanisms (such as the existing Public Advisory Committee) to inform the public of the status of scientific accomplishments and their usefulness in the management of Gulf of Alaska resources. As discussed in Chapter 5, additional ways are needed to increase collaboration between traditional ecological knowledge and modern science. Scientists have learned that traditional knowledge can be a useful source of ecosystem information, for example, the co-management of marine mammals, such as the bowhead whale, by an

Alaskan native commission and federal and state agencies, and the use of Little Diomede Island Inupiat seal-hunting knowledge to capture and track a ringed seal more than 400 miles through the frozen Chuckchi Sea. GEM should foster collaboration with the various Gulf of Alaska communities (see Chapter 5 for community involvement details). Collaboration will advance our understanding of the Gulf of Alaska ecosystem and benefit subsistence and other community resource users.

The GEM Scientific and Technical Advisory Committee, along with interactions with the chief scientist and Program Advisory Committee will need to play a key role in developing the Gulf of Alaska ecosystem monitoring and associated research science plan and in implementing the plan. The Scientific and Technical Advisory Committee in consultation with the chief scientist should provide creative leadership, including the evaluation of GEM's scientific direction; make appropriate scientific program changes when needed; and direct the activities needed to carry out the plan, including solicitation and selection of proposals that best address GEM's goals. Some additional subcommittees may need to be established, and interactions with these could assist the chief scientist and Scientific and Technical Advisory Committee in providing program leadership. Subcommittees should be established, however, only after identification of need. If such committees are arbitrarily established they can be divisive and a hindrance to successful advancement of the program goals.

Proposal solicitations and final recommendations for Trustee Council funding should be a major function of the chief scientist and Scientific and Technical Advisory Committee. The chief scientist and Scientific and Technical Advisory Committee should develop the scientific and technical subjects required to address GEM goals, as well as participate actively in the development of requests for proposals. Workshops hosted by the Scientific and Technical Advisory Committee to determine community-generated research needs may be an effective method for bringing local community resources into the proposal generation and solicitation process. The chief scientist and Scientific and Technical Advisory Committee should organize workshops related to choosing the variables to be monitored over time—keeping in mind that the final selection of variables should be based on hypotheses about how those variables would provide insight into relevant ecosystem processes—and workshops to facilitate the linkage of traditional ecological knowledge with modern science.

If the Scientific and Technical Advisory Committee is to function effectively and play a key role in advising the chief scientist and guiding the GEM scientific and technical program, its membership must be based on their scientific expertise and their ability to translate across the marine habitats and disciplines. Scientific and Technical Advisory Committee members must be perceived to be neutral, unbiased, and focused on the long-term success of the GEM program. The addition of some of its members to the Program Advisory Committee should assist with the integration of local community needs with the GEM scientific research planning process. Scientific and Technical Advisory Committee membership will require regular rotation to obtain the best oversight of GEM over time. Scientific and Technical Advisory Committee members could be compensated and they should have term limits of three to five years, with no direct GEM research or project funding during the period of service.

5

Community Involvement and Traditional Knowledge

Community involvement and the incorporation of traditional knowledge in the Gulf Ecosystem Monitoring (GEM) program is critical to the program's long-term success. Early *Exxon Valdez* Oil Spill Trustee Council documents indicated a desire to incorporate community involvement and traditional knowledge into the new GEM program, and the Trustee Council made many efforts over the past decade to create opportunities for community involvement in the program, with varying degrees of success.

However, this emphasis on community involvement and traditional knowledge appears to have receded in successive documents reviewed by the committee. The committee's interim report discussed the importance of community involvement and use of traditional knowledge and identified a need for increased attention, but the current GEM science plan appears to give these issues less, not more, attention. The committee, once again, urges the Trustee Council to review these issues in earnest. The commitment to and philosophy regarding community involvement and traditional knowledge needs much more clarification and explanation, whether in the GEM plan or in supplementary documents.

The first question to revisit is whether community involvement and traditional knowledge *should be* a part of the GEM program. The committee believes that community involvement and traditional knowledge should be explicitly incorporated in the GEM program. If community involvement and traditional knowledge are to be incorporated, the next question is *why* are community involvement and traditional knowledge important? First, community involvement and traditional knowledge are important because as program components they can contribute to the focus on ecosystem monitoring. Local residents possess valuable ecological knowledge—information that can be directly incorporated into established scientific models. Local residents can be a source of important research questions and can help assure that research is relevant to both ecological and community needs. In addition, local participants offer potential efficiencies in data collection efforts. Local participants are likely to be critical to the success of any stewardship goals associated with the GEM program. Local participation can build constituent support for the GEM program, which is important for a program intended to operate for centuries. Such a partnership has proven successful in Nova Scotia, with the formation of the Fisherman and Scientist Research Society (Box 5-1).

The committee is not alone in recognizing the practical significance of traditional knowledge to contemporary sciences such as ecology, conservation, biology, pharmaceuticals, forestry, fish, and wildlife sciences. The International Union for the Conservation of Nature (IUCN, 1986) lists the following arenas in which traditional knowledge can prove useful to science and environmental applications: new biological insights, resource management, conservation education, reserve design and management, development planning, environmental assessment, and commodity development. Traditional knowledge also has strong potential for informing the science of ecological restoration (Martinez, 1994; Kimmerer, 2000). Ford (2001) suggests that traditional knowledge plays a vital role in ecological monitoring and early warning signs of ecosystem change.

In sum, one answer to the "why" question is that it is in the best interests of the GEM program goals to incorporate community involvement and traditional knowledge. This is a profoundly utilitarian rationale—locals can help the program—but it is potentially a source of foundation for future problems. Such issues should be approached cautiously by the Trustee Council with careful attention given to the cultural and social significance of the participation of the residents of Prince William Sound in the GEM program. Indeed, it appears that the noticeable retreat of communities from GEM program planning activities arises from the perceptions that the relationship between science programs and communities has been relatively one-sided in the past, and that the GEM program will continue this relationship in the future.

The issue of the relationship between the traditional scientific community and the communities of the *Exxon Valdez* oil spill region presents a second broad rationale for incorporation of community involvement and traditional knowledge into the GEM program. The second rationale rests on an equity argument, which is distinct from the utilitarian rationale above. The GEM program, like the Trustee Council itself, is a result of settlement funds dedicated to restoration of an ecosystem damaged by a

> **BOX 5-1**
> **An Example of Community Involvement:**
> **The Fisherman and Scientist Research Society**
>
> Community involvement in scientific research aimed at gaining a better understanding of marine ecosystems can bring benefits. However, communities must have a role in helping to define what will be done and how it will be done. They must be actively involved in conducting the research, analyzing data, and disseminating the results to members of the community and other stakeholders.
>
> One example of community involvement and how long it can take to develop is under way among coastal fishermen and fisheries biologists from the Canadian Department of Fisheries and Oceans in Nova Scotia. The Fisherman and Scientist Research Society was formed in the early 1990s to help develop a common understanding of the status of commercially harvested fishes and invertebrates on the continental shelf off Nova Scotia. Officers of the society are fishermen elected by the membership. The executive is advised by directors at large drawn from the membership and participating member scientists, a Communications Committee, and a Scientific Program Committee. More than 300 members from across the province meet annually to discuss the results of research undertaken in the previous year and to plan major new initiatives. The first several years represented a difficult and uncertain period for the society. It takes time, hard work, and a commitment to succeed to overcome existing biases and to build new relationships based on mutual respect.
>
> Over the past eight years, however, the society has made tremendous strides. It has undertaken collaborative research on a range of topics, including inshore fish abundance surveys, fish tagging, studies on fish diets and physical condition, lobster recruitment, and coastal ocean temperature. The impetus behind most of these studies has come from questions posed by the membership with involvement at the community level. As the society matures the range and scope of the research continues to grow, providing fisheries scientists and oceanographers with an opportunity to address questions that would be difficult to address otherwise.
>
> SOURCE: NRC, 2001.

human technological disaster (Erikson, 1994). This ecosystem includes resource-dependent human communities (Picou and Gill, 1996), and these local communities have strong interest as stakeholders in the outcome of restoration activities (including long-term monitoring). The GEM program is a science program: It can be a science program without the involvement of local people, but it can be fashioned as a science program with effective local involvement with real gains to its relevance and no loss to its scientific credibility.

The equity argument in favor of community involvement compels consideration of some key definitional issues. What do the terms "community" and "involvement" mean? The committee suggests that "community" includes both the geographic communities of the GEM program region and more broadly the people who live and work in that region. Defining "involvement" is more complex and lies at the root of the issues concerning community involvement in the GEM program.

The committee's review of past community involvement in Trustee Council research showed that involvement generally appeared to be a blend of employment opportunities and peripheral advisory roles. The GEM science plan seems to suggest a general continuation of this approach, but with little explanation. However, the committee has received the clear sense that local communities are increasingly uncomfortable with this status quo approach to involvement. It is likely that residents will continue to press for more access to and participation in all phases of the GEM program.

There is abundant literature on traditional knowledge (e.g., Johannes, 1989; Baines and Williams, 1993; Rose, 1993), and on participatory research (e.g., Castellano, 1993; Chambers, 1997; Hall, 1981; Holland and Blackburn, 1998; Park 1993; Park and Williams, 1999). A pervasive theme throughout this literature is the relationship between local people and scientific research programs that is directly relevant to the community involvement/traditional knowledge issues confronting the GEM program. Consider, for example, the distinction between involvement in actual program planning and execution versus providing public advice on programs and projects presented *to* locals, rather than designed *by* locals:

> [T]here is an inherent flaw in calling for more participatory forms of management when the specific goals are predetermined. Under such conditions local people's role in the management process necessarily remains prescribed and largely symbolic. It is the contention of the authors, that whereas there is a discourse of participatory marine management, the practice remains hierarchical and inclined toward use of the knowledge of those with the most formal education and the least experience (Glaesel and Simonitsch, 2001).

Public review does not equal public involvement; it is only part of an overall commitment to public involvement. Similarly, meaningful community participation must consist of more than providing employment to locals (to work on projects conceived and run by others). Seeing local residents only as a potential labor pool ignores the critical factor of *who* asks the research questions. This does not mean that employing local residents is inappropriate, but rather that the continued identification of involvement exclusively with employment is unnecessarily narrow and impedes an understanding of why the relationship between the Trustee Council and local residents is strained.

It might be instructive to consider a reversal of roles. What if the scientific community was treated as a labor pool for a long-term monitoring program administered and controlled by local communities? Can there be any doubt that the scientific community would demand a more substantive role in the program? Of course, either extreme (treating the local communities or the scientific community exclusively as a labor pool and source of secondary advice) is untenable.

If substantive community involvement is to be a feature of the GEM program, the next question is *how* can that involvement be fostered at this planning and initiation stage? Moving beyond mere expression of support for community involvement requires confronting issues of relationships:

> [T]here remains the challenge of establishing effective relationships between the community and external institutions. The power relationships which prevail represent possibly the most critical factor (Castellano, 1993, p. 152).

As we noted in our interim report the entire GEM program needs a foundation that is simple, robust, and adaptable that permits local issues to be addressed in a meaningful way from the very beginning of the program. We noted that there are essentially three possible arrangements to consider in terms of providing a foundation for community involvement. First, every project could be required to feature community involvement. Second, the program could include a separate, distinct community GEM program that would operate with autonomy. Third, the GEM program could be structured to aim for a balanced distribution of power and opportunity between the scientific and local communities.

The first approach is severely flawed because it consists solely of a formulaic insistence on community involvement in every project that will do little more than encourage tokenism. The second approach has merit, but it introduces inevitable difficulties of allocating between communities (or between groups of communities) and would limit opportunities for genuinely mutual exchange between scientists and local residents. The second approach is largely embodied in a proposal put forward by the Chugach Regional Resources Council representing several Alaskan native villages in the GEM region. Alaskan native communities have no direct representation on the Trustee Council and this appears to be a source of tension distinct from more general questions of involvement. The Chugach Council representatives who met with the committee spoke of a desire to institute a community GEM program on a government-to-government basis in terms of their relationship to the Trustee Council. Over the course of the GEM program it appears that the Trustee Council will have to be sensitive to sovereignty issues regardless of whatever actions are taken in terms of incorporating Alaskan native involvement in the GEM program.

The committee repeats its recommendation from its interim report: GEM should pursue an approach to community involvement based on shared power and shared opportunity between the scientific and local communities. The goal of shared power requires community representation at all organizational levels. For community-originated studies to be effective these structural provisions of power to communities must be accompanied by opportunities to receive funding. To ensure genuine incorporation of community interests and local knowledge and experience, the program should have some flexibility to fund proposals written outside the standard format and phrasing of the scientific establishment. There might also be a mechanism (e.g., periodic training sessions) to support communities wishing to submit proposals.

The institutional and communicative barriers confronting communities can be substantial. For example, Castellano (1993) states:

> [C]ommunity groups typically encounter resistance in local and regional agencies to community-sponsored proposals to vary the application of inappropriate rules.
>
> A second issue is management of communications between communities and institutions when the actors operate from differing styles of communication. In general, the greater the distance between the cultural forms prevalent in the community and the cultural forms recognized or legitimated in the institutions, the more difficult it will be for both sides to recognize the commonalities that permit accommodation of community proposals by the institutions. If congruence between community proposals and institutional priorities is not easily identified, advocates within the institution will be subjected to personal risk in attempting to sell the ideas to their colleagues. The packaging of community proposals to emphasize points of congruence between new approaches and accepted practices, and the identification of persons or units in the institutions with a mandate to act in the field are strategic imperatives (Castellano, 1993, p. 153).

The kinds of barriers to effective community involvement highlighted in the literature are evident in the GEM planning process. For example, the committee was informed that one significant aspect of community involvement envisioned for the GEM program consisted of the subcommittees featured in the discussion of "guidance on GEM program development and implementation" in Section 6.3 of Volume I. The description of the subcommittees (p. 70) underscores some of the communicative and perceptual challenges confronting program planners and local communities.

> *The subcommittee would be composed of scientists, resource managers, and other experts selected primarily for disciplinary expertise and familiarity with the broad habitat type (watersheds, intertidal and subtidal, ACC, and offshore). Institutional and professional affiliations would be of interest in selecting members to promote collaboration and cooperation.*

The essence of the problem here is that the very language that is ostensibly intended to invite community participation is instead likely to be interpreted as repelling community participation.

In summary, the committee recommends that community involvement be designed throughout the GEM program in a manner that promotes meaningful involvement and provides for flexibility into the future as the GEM program evolves. Approaching community involvement in the fashion recommended by the committee should be regarded as a work in progress, because building the necessary relationships and developing a process that works will take time (see Box 5-1). In many respects the GEM program will be breaking new ground in integrating community involvement into a long-term science plan. As one step in rethinking its commitment to community involvement, the Trustee Council should review community outreach programs designed by the Prince William Sound Regional Citizen's Advisory Council, which have been successfully used in communities and native villages affected by the *Exxon Valdez* oil spill (<www.pwsrcac.org>). This may provide direction for designing activities that promote substantive participation and involvement of local residents in all phases of the GEM program.

The committee is under no illusion that successful incorporation of community involvement and traditional knowledge in the GEM program will be easy. It will take more than just the inclusion of the words "community involvement" and "traditional knowledge" in program planning documents. It will require the engagement of planners, administrators, and researchers representing the scientific community with relevant experts and literature regarding participatory research and traditional knowledge, and most of all, with residents of local communities on shared terms. It will require the local communities to recognize that the GEM program will not address all their needs and aspirations. Nonetheless, the opportunity to develop community participation in the GEM science program will benefit all parties involved and should be seriously pursued by the Trustee Council.

6

Data and Information Management

Efficient archiving and dissemination of data is critical to any long-term research program. Careful, early attention to data management can ensure that the data collected are truly useful in capturing trends and illustrating changes in the system over time. The Long-Term Ecological Research sites supported by the National Science Foundation again provide models of how to organize and manage long-term ecological data sets. The GEM program must include a strong commitment to data and information management. To extract the full scientific value of GEM, data and information must be made available to the scientific community, resource managers, policy makers and the public on a timely basis. Data management must be designed to facilitate data exchange among GEM scientific investigators, make data available to the public and outside scientific community, and archive the data products.

The success of GEM will be critically dependent on establishing some kind of Data Management Office, which would be staffed with a data manager and others as needed to organize, disseminate, and archive the data. The data manager would participate in the planning of the sampling program, organizing the data, assuring data quality, archiving the data and providing data to the principal investigator and public. There should be a Data Management Subcommittee to help provide periodic outside advice on data policies; the data management system; preservation of data with relevant documentation and metadata; advice on enforcement of data policies; and to facilitate exchange of data with related oceanographic programs. Both data managers and scientists should serve on the Data Management Subcommittee to facilitate the interaction of scientists with the data management staff so that data management policies and procedures are in tune with the scientific focus of GEM. These groups would develop a data policy that establishes the rules for submitting data and models; facilitates quality control of the data by the data management office; ensures that the data are properly archived; ensures the rights of the scientific investigators; promotes the exchange of data between investigators; and ultimately, makes the data available to the general public and outside scientific community. These data management policies are followed by large scientific oceanographic programs such as the Joint Global Ocean Flux program (<usjgofs.whoi.edu>), Global Ecosystem Dynamics (<globec.oce.orst.edu/groups/nep>), and the Coastal Ocean Processes program (<www.skio.peachnet.edu/coop>).

GEM needs to be committed to the timely submission and sharing of all data collected by its researchers. In accepting support each principal investigator should be obligated to meet the requirements of the GEM data policy. These should include submitting collected data in the established format within set periods from collection. Investigators should be encouraged to exchange data and models with other GEM scientists to promote integration and synthesis.

Data management must have sufficient resources to accomplish its necessary functions in support of the GEM program. According to recent reviews, some of the most successful coastal monitoring efforts allocate as much as 20 percent of their total budget toward data management (Sustainable Biosphere Initiative, 1996; Weisberg et al., 2000). To be successful GEM will need to make a similar financial commitment to data management. A program such as GEM with a long commitment to observations of ecosystem processes will be viewed regionally, nationally, and internationally for leadership in data management.

A body of data exists for the Gulf of Alaska to which GEM investigators will need ready access. One of the first tasks of the Data Management Office should be to install this relevant data into the GEM database. Examples of pertinent ancillary data sets are NOAA's Tropical Atmosphere-Ocean El Niño Southern Oscillation data, Pacific Decadal Oscillation estimates, the Gulf of Alaska Global Ecosystem Dynamics program, and historical regional oceanographic and climate data. Another example is the North Pacific Marine Science Organization's Technical Committee on Data Exchange Website that contains links to long-term, interdisciplinary data sets for the North Pacific. These data archives will be essential to ecosystem modeling and synthesis in the GEM program. Also essential to the initial planning of the GEM program will be data collected in the past decade with *Exxon Valdez* oil spill funding. These data need to be synthesized to guide the selection of the sampling sites and measured parameters of the GEM coastal time-series observations. These data must also be made available to collaborating scientists, scientists outside the program, the public, and resource managers.

The policy of such federal agencies as the National Science Foundation, Office of Naval Research, and the National Oceanographic and Atmospheric Administration is that two years after collection, data should be available to the general public and scientific community through the National Oceanographic Data Center (NODC). Data collected by the GEM program should be submitted to the NODC in addition to being made available to the public through the GEM Website or similar structures.

The general description of the data management architecture in the draft GEM science plan is very good. The data management functions of data receipt, quality control, storage and maintenance, archiving, and retrieval are recognized and adequately addressed. The report recognizes that different types of data products will be needed for basic research and analysis, modeling, resource management applications, and public outreach. Access to the data archives and software display will be an important public outreach component. There would be multiple levels of complexity to the data access ranging from users with limited backgrounds with these data to use by the investigators who gathered the data.

One of our chief concerns is the importance of having clear, established data policy and a willingness to enforce it. One of the first tasks of the GEM Data Management Subcommittee should be to establish a data policy to which all investigators must adhere and to help GEM set up the structure of the Data Management Office. It was apparent in reviewing the *Exxon Valdez* Oil Spill Website that it was difficult or impossible to retrieve data collected from past research projects. This trend must change if the GEM program hopes to realize its potential for understanding the Gulf of Alaska ecosystem. Data collected should be easily retrieved by various user groups, as is the case for programs such as the Joint Global Ocean Flux Experiment (<www.usjgofs.whoi.edu>), Global Ocean Ecosystem Dynamics Experiment (<globec.whoi.edu and globec.oce.orst.edu>), or, more generally, the data available from the National Snow and Ice Data Center (<http://nsidc.org/index.html>). The Data Management Office must have sufficient staff and infrastructure support for receipt, quality control, archiving, and retrieval of data products required by its upser groups.

7

Synthesis, Modeling, and Evaluation

Writing a science plan to guide the Gulf Ecosystem Monitoring (GEM) program for the next 100 years is no easy task. It is simply not possible to know everything that should be addressed. To be useful over the long term, the plan will need to be flexible. The issues in 10 years, or 20, or 50 may be different from today's issues. Concerns about the ecosystem may change in the face of the possibility of increased tourism, terrestrial resource harvests (timber), hydroelectric development, and other changes in water usage and land use. Even so, we must qualify that we do not expect the GEM document to address each of these issues. This is where flexibility becomes important. The plan needs a system in place for synthesis of knowledge at specific points in time and evaluation of what has been learned and what needs to be done next to progress in understanding the ecosystem.

SYNTHESIS

An initial synthesis needs to include several components. The first step, a much-needed literature review, has been completed in the "Scientific Background" section in Volume II, Part 3, of the GEM plan (EVOSTC, 2001). Recent information from other geographic areas that contain relevant information can be incorporated when needed for specific topics. The second step, compilation, assessment and analyses of databases, has not been done. This step is critical to accommodate the imperative third step, which is a synthesis of *Exxon Valdez* oil spill research from 1989 to the present. Though a few programs have completed synthetic views of their results (e.g., *Fisheries Oceanography* vol. 10, [Suppl. 1] – "A Sound Ecosystem Assessment [SEA] Synthesis"), most have not. Many studies that have been funded over the past 13 years have yet to be published. Annual reports are not publications and certainly do not qualify as syntheses.

The knowledge gained about Prince William Sound is extensive because of *Exxon Valdez* oil spill funding. Retrospective analyses have led to new hypotheses and ideas in many instances, not the least of which is the concept of a "regime shift" (Francis and Hare, 1994; Hollowed and Wooster, 1995; Anderson and Piatt, 1999) and the Pacific Decadal Oscillation (Mantua and Hare, in press). However, there is much more to be gained from past studies that should be used to direct the future of GEM. The completion of the third step will lead to the fourth step: assessment of accomplishment of past goals. The synthesis of data and assessment of what has been learned in the recent studies will provide a starting place from which to hone hypotheses needed to direct GEM research.

The generation of new hypotheses will lead to proposals for new work, which in turn will lead to the need for additional synthesis. Synthesis is an iterative process and as such is both the first and last steps. For GEM to continue to be successful, periodic re-synthesis of new data will be needed. A synthesis will assure that there is not a long lag time in publication of results and access to data of other GEM researchers, such as currently experienced under *Exxon Valdez* oil spill. A periodic synthesis on the scale of the five-year increments will promote comparisons between past and recent conditions. Additionally, scheduled syntheses will ensure evaluation of program direction.

> *One presumption in a long-term program is that technology will change, providing opportunities for collecting new data types or collecting existing data more efficiently. Another presumption is that users will become more sophisticated, and their needs will change as they become accustomed to the data streams that are produced. Many successful programs incorporate periodic program review to assess how the program should change in response to these new collection opportunities and needs.* (Weisberg et al., 2000).

The synthesis will tell whether the science plan and the structure of the program is working.

As GEM is envisioned to be a 100-year plan, we suggest that a time line on a scale longer than five years be included in the GEM plan. We have emphasized that long-term research is the linchpin of this program, and the projected time line should reflect that effort. Within that time line periodic syntheses should figure prominently. Synthesis should be viewed as a key component of the plan and funding for synthesis should be incorporated. While periodic review is necessary, the long-term research should be modified only when a strong case can be made for improving the program (Weisberg et al., 2000). The synthesis and review should involve a wide range of scientists and community members, as data users are critical to the review process (Weisberg et al., 2000).

MODELING

Synthesis and modeling are interconnected. For example, one first could create a conceptual model that will tell which quantities need to be measured, collect data, synthesize data, and then create a more quantitative model. Alternatively, one could collect and synthesize data, and then create a statistical model that could be used to collect more data to verify the model. In a third approach, one could perform a synthesis on retrospective data and create a working model, also known as an hypothesis, which would be used to design data collections that are synthesized into more sophisticated models. Note that the models and syntheses may take many forms from conceptual to highly quantitative. Regardless of the order of these steps and the sophistication of the techniques, the components of synthesis and modeling are both critical. The combination of synthesis and modeling are tools for evaluation of past work: testing the appropriateness and accuracy of hypotheses and generation of new hypotheses. This approach will keep the GEM program moving forward by addressing issues that arise from the conceptual foundation and filling gaps identified during the evaluative process.

The elements of a successful modeling component are outlined in the GEM monitoring plan. It is worth emphasizing that modeling should be a component in all phases of GEM as a research, synthetic, and diagnostic tool. The strategic elements for a successful ocean-observing program are a combination of in situ observations, remote sensing, and modeling (IOC, 2000). All three elements complement each other to provide a more comprehensive view of the environment. Because of the different spatial and temporal scales of response and variability in the physical environment and living resources of the Gulf of Alaska, models will be needed to merge disparate and discontinuous measurements. A hierarchy of models (statistical, theoretical, empirical) should be employed in the GEM program. The skill of models should be routinely assessed. Some models will require some form of data assimilation using information collected during the monitoring program. The data are inserted into the model to ensure that the model outcome more closely resembles the in situ observations. The GEM program should work toward more realistic and accurate numerical models for the prediction of ecological processes. The unparalleled opportunity of a long-term observation program in the Gulf of

Alaska coupled with a concerted effort in modeling will produce exciting new tools for the management of the Gulf of Alaska's living resources.

REVIEW OF THE GEM SCIENCE BACKGROUND SECTION

GEM planners have already made a first synthesis by compiling information in the GEM planning document (EVOSTC, 2001). The current "Science Background" section is a good comprehensive review of relevant knowledge. The document establishes a common background that can be used as source material. This should stand as an indication of what is known at this time. This state of knowledge in this work plan does not need to be updated, as the updating will take place routinely through GEM synthesis efforts. This is an excellent background from which synthesis efforts can begin.

We applaud the GEM writing committee on the excellent scientific background that they created in Volume II, Part 3. This scientific background contains up-to-date knowledge and is well presented. In most cases there is a referenced, accepted scientific basis for the material presented. The use of figures to demonstrate concepts and points is well done. This document will be useful to inform the Trustees, scientific community, and the public. We recognize, however, that all interested parties will not read the entire document; we suggest that the "Executive Summary" highlights in non-technical language the main scientific points on which GEM is based.

Generally the physical oceanography is well presented in Volume II of the GEM document. The major deficiency is the lack of attention to processes that might take place on the mid-shelf. While the shelf is addressed in the document, when the choice of habitats is selected, the document turns rather quickly from the Alaska Coastal Current to the offshore areas of the shelf break, continental slope, and deep ocean basin. The mid-shelf region might be very important to the nutrient fluxes and primary production of the region, because relatively deep nutrients must get into the euphotic zone, and the pathway is unknown.

There are some smaller inaccuracies and over-simplifications in the physical oceanography section. For example, the definition of the shelf as being located at depths of less than or equal to 200 m is wrong, given that there are many locations deeper than that, including locations in Prince William Sound. There are also some problems with the discussion of circulation in Prince William Sound. Although this circulation is intimately connected with the circulation of the Gulf of Alaska, the plan emphasizes the circulation of the central Gulf of Alaska over the circulation over the adjacent shelf, and the thrust of this document pushes the studies into the deep Gulf of Alaska.

In the GEM plan the discussion of time and special scales is very brief. This topic might well be the weakest part of the GEM program. The processes that affect primary production are going to have space scales on the order of kilometers. Single monitoring stations will not be useful tools. Granted, Ocean Station P and GAK1 measurements have added to our understanding of the system, but these are really "first looks" similar to an initial Mars probe. From ongoing studies, mesoscale physical and biological processes on the shelf are appearing to be important in the Gulf of Alaska. A program to measure on these time and space scales over the entire shelf will be very, very expensive to maintain. In addition, it is important to make measurements in winter, as this might well be the most critical time for the marine populations. Or GEM could break the problem; for example, in meteorology the long period changes are climate-related problems whereas there are daily changes (weather) embedded in these long-term processes. There are similar time and space scales in oceanographic processes, and sampling must be designed to measure all these scales. There is no distinction in the document with regard to the atmosphere. For example, GEM should develop studies to address the seasonal variability embedded in the long-term monitoring program. Three to five years of seasonal measurements will be required to determine the seasonal signal. After those studies scientists should be able to reduce the measurements into a monitoring mode, assuming that an increased understanding will allow more targeted sampling. Unfortunately, there is no example of a system in which this has been done.

There are some physical science statements with which we disagree or question. We question the source of the statement about long-term warming of the northeastern Pacific Ocean. This has not been substantiated with data to date. The longest air temperature time-series for the region (Sitka, Alaska) shows no increasing trend since 1828 (Royer, 1993). We question where the iron limitation hypothesis came from. The hypothesis that the primary productivity on the shelf of the northern Gulf of Alaska is not documented. It seems likely that there is enough iron from terrestrial sources to offset any depletion, however, these measurements have not been made.

The biological support for the science is good, and we commend the GEM team for this strong compilation of the current state of knowledge. Simultaneously, we would like the GEM plan to recognize the tentative nature of some of the most recent unpublished findings. Be aware that the conclusions may change when studies are completed and prior to publication. GEM should not be dependent on tentative findings.

A 100-year plan should be only a broad outline with details to be worked out in work plans. A broad-brush understanding of the area in question at this time in history is necessary for the start of a 100-year plan.

It is inappropriate to include detailed research questions in the "Scientific Background" section, such as: "Do diurnal-period shelf waves along the Kodiak shelf influence biological production and the dispersal of planktonic organisms (EVOSTC, 2001, Vol. II, p. 64)?" We suggest that these questions be removed from the document. The objective of this section of the document is to set the stage for the scientific questions and hypotheses to be generated. We cannot fault the questions themselves, because they ask just about everything. They are at once extremely general and too detailed. Including this level of detailed questions in the background of this document leads us as reviewers to believe that all research will be restricted to addressing these specific questions. That would discourage original hypothesis generation and research in the proposal process.

In conclusion, we believe that the GEM plan we reviewed provides an excellent scientific background for the Gulf of Alaska region. We want to see a synthesis of data that have been collected under *Exxon Valdez* oil spill and we want to see periodic re-synthesis and evaluation. We suggest that various types of modeling will be useful tools to aid this synthetic process.

8

Conclusions and Recommendations

The *Exxon Valdez* Oil Spill Trustee Council is to be commended for its foresight in setting aside funds over the years to create the trust fund to provide long-term funding to the Gulf Ecosystem Monitoring (GEM) program. The GEM program will offer an unparalleled opportunity to increase understanding of how large marine ecosystems in general and Prince William Sound and the Gulf of Alaska in particular function and change over time. The committee believes this program has the potential to make substantial contributions of importance to Alaska, the nation, and environmental science.

Since this committee was formed in June 2000, it has met five times to learn about and discuss the GEM program. We have conveyed our comments and recommendations in a letter report (November 2000) with advice on program timing and a more detailed interim report (February 2001) that critiqued an early draft of the program science plan. These reports focused on the early planning, were specific to the draft planning documents, and were primarily directed to program staff. In this final report we provide broader comments and a document that has more general and far-reaching lessons about which elements are essential to the success of a long-term research and environmental monitoring program such as GEM.

GEM's mission as stated in EVOSTC (2000a), is ambitious: "to sustain a healthy and biologically diverse marine ecosystem in the northern Gulf of Alaska and the human use of the marine resources in that ecosystem through greater understanding of how its productivity is influenced by natural changes and human activities." The purpose of any mission statement is to serve as a general guiding principle and statement of underlying philosophy and approach, and this mission statement accomplishes this purpose. However, putting this statement into practice is likely to prove difficult.

According to an early EVOSTC document (EVOSTC, 2000b), GEM was conceived to have three main components:

1. long-term ecosystem monitoring (decades in duration);
2. short-term focused research (one to several years in length); and
3. ongoing community involvement, including use of traditional knowledge and local stewardship.

The committee still views this early vision of the program as a sound foundation on which to build. In a later document (EVOSTC, 2000a) the purpose of the GEM program is further delineated to contain five program goals: detect, understand, predict, inform, and solve. The committee understands the general intent of these goals and the necessity of making the program respond to both the needs of science and the needs of its political constituency. But as discussed in earlier reports, the committee remains concerned that these five goals are extremely diverse and far-reaching. While the GEM mission is a good general statement of intent, the committee's concern is that addressing all five goals will present the risk that the research and monitoring program will be spread too thin to be effective.

In its review of the evolving GEM long-term research program the committee noted some positive strides. We believe that the GEM planners tried to include the interests of diverse stakeholders (Trustee Council, scientists, various advisory groups). We are pleased to see that the planning process has caused an evolution in the draft and the thinking behind it. We commend GEM planners for not taking the easy route of simply picking stations and starting data collection, and that they took the time to think about the conceptual foundation and develop the hypotheses that are necessary to define data needs. We find the conceptual foundation is much improved; however, placing the conceptual foundation deep in Volume II of the plan is not appropriate. That late placement implies that it is an afterthought and not the foundation upon which the program is built. It is, however, a good point of departure for GEM, and we assume it will evolve as the program moves toward implementation. We believe that GEM planners have made progress on the development of hypotheses, although there is still room for more work in this area.

GEM staff members have made a good effort to reach out to the science community. They have a good start on their discussion of and approach for using modeling effectively; and they have made very good progress in setting up a strategy for data management. We found that the science review section is very useful. Although it may seem obvious, many of these positive strides have occurred because the Trustee Council and GEM staff have set up a planning process and are allowing time for the evolution of thinking.

The committee has struggled, however, with its basic charge (to review the GEM program) because the program was literally evolving as we worked and we often were dealing with a "moving target." We also struggled because, as scientists, we are more accustomed to dealing with research programs instigated and directed by scientists, such as the Global Ecosystem Dynamics program, or by agencies with clear mandates, such as Mineral Management Service's Environmental Studies program. Instead, GEM is a research program directed by a Trustee Council made up of six agency representatives, each carrying responsibilities for mission-oriented state and federal agencies. Their role is made especially difficult because of the legal requirement that all their decisions be unanimous. GEM is supported by a staff that includes both scientists and non-scientists who have the unenviable job of balancing not only the expectations of the science community (the norm when developing a new science program) but also the expectations of various other Alaskan stakeholders and the inevitable political forces of the Trustee Council itself.

While this committee whole-heartedly endorses the idea of a long-term ecological research program in the Gulf of Alaska and commends the Trustee Council and other decision makers for creating such a program, we must stress that this report is not an endorsement for implementation of the GEM program as currently designed. Our proposed changes are described in the following conclusions and recommendations.

CONCLUSIONS AND RECOMMENDATIONS

Opportunity for Sustained Study

Conclusion: GEM is an important opportunity to do truly long-term research in a marine ecosystem, and this long-term approach is essential to distinguish natural variability from human impacts. The long-term nature of the program, intended to cover a period of many decades, is the flagship contribution of the plan. Long-term monitoring by definition must include sustained, consistent observations over a long period and thus requires a long-term commitment from the highest levels of decision makers. This commitment will require a substantial financial investment.

Short- and medium-term research is an appropriate way to address current questions and management needs, but the fundamental importance of the long-term program should not be lost.

Recommendation: The majority of GEM funds should be spent on long-term monitoring and research, that is, sustained observations of ecosystem components and ecological processes over decades. This long-term perspective will be the GEM program's special contribution to scientific understanding in Alaska's marine environment; most other research programs are short-term. These long-term measurements will be necessary to differentiate the effects of natural variation from human-induced changes on the Gulf of Alaska ecosystem. The coastal Long-Term Ecological Research sites funded by the National Science Foundation provide good models of such long-term research.

Elements of a Sound Long-Term Research Plan

Conclusion: A sound, long-term research plan must clearly define its conceptual foundation, scope, organizational structure, data management methods, and methods for periodic synthesis and review. The conceptual foundation presented in the draft science plan is adequate and with modest restatement as a hypothesis could be a useful focus for research. The science plan and research objectives need to be directly linked to this conceptual foundation.

Recommendation: The current draft science plan (EVOSTC, 2001) needs to be shortened considerably by removing tangential materials so that it is a clear guide for the future. The conceptual foundation needs to be discussed early in the GEM planning document because that placement captures its importance as the fundamental building block on which the rest of the program depends. The science plan should include a broad conceptual foundation that is ecosystem-based. It should seek to understand natural and human-induced changes and it should be flexible to accommodate changing needs without compromising core long-term measurements. These hypotheses will provide a bridge between the conceptual foundation and the eventual implementation of the science program. Because the conceptual foundation states that the ecosystem is affected by both natural variability and human-induced change, as the plan is implemented both of these drivers should be addressed in studies.

Implementation of the GEM Program

Conclusion: The planning process for GEM has been difficult and costly, but the investment in planning is critical for success. Long-term measurements cannot begin until after the appropriate variables have been identified, and these must be based on the conceptual foundation and hypotheses. The planning and design of sampling will continue to take considerable time and effort in the early years of the program. It is more important to identify the right variables than to rush to collect data.

Recommendation: The GEM plan and planning process need to provide careful consideration of what to measure, how often, and where, based on input from a broad cross-section of the scientific community, local communities, and managers. These decisions on hypotheses and attendant measurements should be made by the chief scientist working with the Scientific and Technical Advisory Committee and other independent scientists and stakeholders over the course of several years as program implementation gets under way.

GEM's Role in Gulf of Alaska Research

Conclusion: GEM's primary goal should be to develop a comprehensive and eventually predictive understanding of the Gulf of Alaska ecosystem. The long-term nature of GEM will enable it to serve as a framework for marine research in the Gulf of Alaska. Other programs will come and go on shorter time frames and should be encouraged to coordinate with GEM, but GEM does not have the resources to be the central coordinating body for all such efforts.

Recommendation: The focus of GEM should be its long-term program, and GEM decision makers should not try to do too much or this will dilute GEM's limited resources and impact. Because of the long time frame of GEM, it can provide a building block for partnering with other programs that will come and go, but it should not be distracted by the idea of assuming leadership of Gulf of Alaska marine research.

Recommendation: GEM should not see its role as filling the gaps in other programs, because adding these kinds of activities will inevitably erode funding for the GEM core measurements. This does not preclude GEM from involvement in other programs in which the research is addressing issues or collecting data that has been identified as necessary for addressing the central hypotheses of GEM.

Recommendation: It simply is not possible for GEM, given its resources, to play a leadership role in both scientific research and day-to-day support of resource management. GEM should not be involved in the types of monitoring that are typically the responsibilities of agencies. GEM should not subsume routine surveys, stock assessments, and data collection that have been the normal province of resource management agencies. Of course, a large monitoring program like GEM will supply much information that is useful to resource management agencies as a result of its own activities.

Community Involvement

Conclusion: The GEM plan does not currently describe effective and meaningful ways to involve local communities. This involvement should occur at all stages, from planning (e.g., selecting the questions to be addressed and variables to be monitored) to oversight and review. Local knowledge and traditional ecological knowledge can be used to generate ecologically sound and socially relevant research ideas. Science and community partnerships can lead to achievements that neither could attain independently. Specifically, such collaborations provide scientific knowledge as well as community education and local support of science. These outcomes are important especially because of the long-term nature of GEM; such involvement might be less critical in shorter programs, but the century scale requires the establishment of long-term bonds.

Recommendation: The Trustee Council and GEM program staff must continue to seek ways to build meaningful community involvement at all stages of planning and implementation, from selecting the questions to be addressed and identifying the variables to be monitored to providing program oversight. It was outside the scope of this committee to advise specifically on what programs or methods to use; neither are we as experienced as GEM staff in dealing with Alaska's diverse communities of interest. Nonetheless, we are certain that the community involvement debate will continue until better resolution of this issue is found.

Geographic Scope

Conclusion: No program can be expected to meet the needs of all potential data users, and tradeoffs are inevitable between the intensity and spatial range of sampling. That is, if the scope of GEM is physically large, then its long-term research component will be able to collect less information at any one site (because there is a finite amount of information that can be collected with finite financial resources). If the scope of GEM is physically smaller, there can be more monitoring sites or more types of information collected. Research projects and sampling will need to be selected very carefully to avoid diluting activities so that their usefulness is limited. GEM planners can choose to obtain more limited information from a large area or more in-depth information from a smaller area.

Recommendation: GEM planners must make an explicit choice on how to focus the program's research. There are many options for carrying out coordinated research that avoids piecemeal projects. One option is to concentrate on a particular geographic area, as the committee recommended in its interim report. Another possibility is to target a few variables across a broad geographic range, such as measuring physical oceanographic variables over long time periods (temperature, salinity, currents). It is possible to concentrate attention on particular habitats in a large geographic range. These choices must be guided by the conceptual foundation and the hypotheses selected for investigation.

Using Habitat as an Organizing Concept

Conclusion: GEM or any large research program can organize its effort and funds in many ways and still be successful. The habitat approach described in the GEM science plan is one way of dividing attention and funds, and it has the advantage of being understandable to many of the program's key stakeholders. GEM planners need to be aware of its one critical disadvantage: a habitat approach can fail to address key linkages, flows, and processes between habitats, which is where many of the most interesting lessons of the long-term GEM program might be seen.

Recommendation: Given the habitat approach selected GEM planners must make a concerted effort to ensure that the program has clear, concrete mechanisms to address cross-habitat links. This does not necessarily mean creating a linkage subcommittee but rather building into each habitat study the opportunity to make measurements of flows among habitats and highlight other interactions. Across-habitat connections must be addressed during synthesis and modeling. These efforts are essential to creating a truly integrated program, where the whole is greater than the sum of the parts.

Organizational Structure

Conclusion: The GEM research plan is being developed to carry out long-term research, short-term research, and synthesis and modeling of data sets. Soliciting proposals, evaluating proposals, and the time frame for the research effort and its funding will differ for these scientific activities. The current science plan does not distinguish among these activities in terms of the procedures necessary to manage them and achieve useful results, or even that the goals of these three approaches differ. Strong scientific guidance is required through all the activities of GEM.

Recommendation: GEM planners, with input from the science community, should identify how these three kinds of scientific endeavors will be incorporated and managed within the science plan. For instance, long-term research projects, short-term research projects, and synthesis efforts will require different mechanisms for proposal solicitation and evaluation and different time frames for funding.

Recommendation: The scientific leadership of the GEM program should be in the hands of a chief scientist advised by a Scientific and Technical Advisory Committee. The chief scientist should have adequate assistance to execute the program.

Conclusion: The organizational structure supporting GEM needs to ensure ongoing, independent scientific oversight and review. It should be easy for new researchers and local community members to be involved in planning and carrying out the research projects. If the Scientific and Technical Advisory Committee is to function effectively and play a leadership role in developing and directing the GEM scientific and technical program, its membership must be selected carefully.

Recommendation: The Scientific and Technical Advisory Committee will play a key role in leading the GEM program and ensuring program credibility. Scientific and Technical Advisory Committee members should be chosen based on their scientific expertise and their ability to link across the marine habitats and disciplines. To obtain the best program oversight over time there should be regular rotation of the members of all advisory groups, such as the Scientific and Technical Advisory Committee. Advisory Committee members should be and should be perceived to be neutral parties who are focused on the long-term success of the program. Members may need to be compensated for their service; they should have term limits of three to five years with no direct GEM research funding during their period of service.

Recommendation: The design of proposal solicitations and final recommendations for Trustee Council funding should be major functions of the Scientific and Technical Advisory Committee and chief scientist. In designing proposal solicitations, the Advisory Committee should be responsible for developing the scientific and technical subjects required to address GEM goals. Community workshops hosted by the Scientific and Technical Advisory Committee would be one method to help articulate community-generated research needs and could be a way to increase the participation of local communities that use Gulf of Alaska resources. The Scientific and Technical Advisory Committee and chief scientist should be responsible for organizing workshops designed to provide input on core variables to be measured over time. Final decisions on variable selection can be based on hypotheses proposing how each variable provides insight into human and climate-based changes in the ecosystem.

Recommendation: There should be an open process for nominating individuals to serve on the Scientific and Technical Advisory Committee, both during its initial formation and as the GEM program continues. Various independent scientific groups can assist in the initial formation to help broaden the selection process and find candidates with suitable experience in the initiation and implementation of large-scale, long-term ecological research. The chief scientist should review the nominations and recommend selections, with appropriate documentation, to the Trustees, who are responsible for the appointments.

Data and Information Management

Conclusion: There will be significant costs associated with data and sample processing and with data archiving. It is a common mistake to underestimate the cost of data and information management. To extract the full scientific value of any research program data and information must be made available to the scientific community, resource managers, policy makers, and the public on a timely basis. Each of these audiences will require information in a different format. The committee commends the initial development of data management procedures; careful implementation of these procedures is key.

Recommendation: GEM should create a comprehensive Data Management Office (not just an archive but a group of people who address these issues). Other large science programs spend as much as 20 percent of funds on data management. The multi-decadal scale of GEM will require a similar commitment.

References

Anderson, P.J., and J.F. Piatt. 1999. Community reorganization in the Gulf of Alaska following ocean climate regime shift. *Marine Ecology Progress Series* 189:117-123.

Baines, G., and N. Williams. 1993. Partnerships in tradition and science. Pp. 1-12 in N. Williams and G. Baines (eds.), *Traditional Ecological Knowledge: Wisdom for Sustainable Development.* Centre for Resource and Environmental Studies, Australian National University, Canberra.

Carpenter, S.R., J.F. Kitchell, and J.R. Hodgson. 1985. Cascading trophic interactions and lake productivity. *BioScience* 35:634-639.

Castellano, M.B. 1993. Aboriginal organizations in Canada: Integrating participatory research. Pp. 145-155 in P. Park, M. Brydon-Miller, B. Hall, and T. Jackson (eds.), *Voices of Change: Participatory Research in the United States and Canada.* Bergin & Garvey, London.

Chambers, R. 1997. *Whose Reality Counts: Putting the First Last.* Intermediate Technology Publications, London.

Driskell, W.B., J.L. Ruesink, D.C. Lees, J.P. Houghton, and S.C. Lindstrom. 2001. Long-term signal of disturbance: Fucus gardneri after the Exxon Valdez oil spill. *Ecological Applications* 11(3):815-827.

Erikson, K. 1994. *A New Species of Trouble: Explorations in Disasters, Trauma, and Community.* W.W. Norton, New York.

EVOSTC (*Exxon Valdez* Oil Spill Trustee Council). 2001. *Gulf of Alaska Ecosystem Monitoring and Research Program (GEM).* NRC Review Draft. Volume I: Strategic Plan for Monitoring and Research. Volume II: The Historical Legacy: Building Blocks for the Future. Volumes I and II together are referred to in this report as the GEM Science Plan.

EVOSTC (*Exxon Valdez* Oil Spill Trustee Council). 2000a. *Gulf Ecosystem Monitoring: A Sentinal Monitoring Program for the Conservation of the Natural Resources of the Northern Gulf of Alaska.* GEM Science Program NRC Review Draft April 21, 2000.

EVOSTC (*Exxon Valdez* Oil Spill Trustee Council). 2000b. Can we predict the ways of the sea? *Restoration Update* 7(1):2, 10-11.

Ford, J. 2001. The relevance of indigenous knowledge to contemporary sustainability. *Northwest Science* 7:185-190.

Francis, R.C., and S.R. Hare. 1994. Decadal-scale regime shifts in the large marine ecosystems of the north-east Pacific: A case for historical science. *Fisheries Oceanography* 3:279-291.

Glaesel, H., and M. Simonitsch. 2001. The Discourse of Participatory Democracy in Marine Fisheries Management. Paper presented at the Putting Fishers' Knowledge to Work Conference, University of British Columbia, Fisheries Centre, August 27-30, 2001.

Hall, B. 1981. Participatory research. Popular Knowledge and Power: A personal reflection. *Convergence* 14:6-19.

Holland, J., and J. Blackburn (eds.). 1998. *Whose Voice: Participatory Research and Policy Change*. Intermediate Technology Publications, London.

Hollowed, A.B., and W.S. Wooster. 1995. Decadal-scale variations in the eastern subarctic Pacific. II. Response of northeast Pacific fish stocks. Pp. 373-385 in R.J. Beamish (ed.), *Climate Change and Northern Fish Populations*. Canadian Special Publication of Fisheries and Aquatic Sciences, No. 121. National Research Council of Canada, Ottawa.

Hood, D. W. 1986. Physical Setting and Scientific History. Pp. 5-27 in D.W. Hood and S.T. Zimmerman (eds.), *The Gulf of Alaska: Physical Environment and Biological Resources*. MMS publication number MMS86-0095. U.S. Department of Commerce, National Oceanic and Atmospheric Administration, and Minerals Management Service, Washington, D.C.

IOC (Intergovernmental Oceanographic Commission). 2000. Strategic Design Plan for the Coastal Component of the Global Ocean Observing System. October 2000. GOOS Report No. 90. 99 pp + 6 annexes.

IUCN (International Union for Conservation of Nature) 1986. Tradition, Conservation and Development. Occasional newsletter of the Commission on Ecology's Working Group on Traditional Ecological Knowledge. No. 4. International Union for Conservation of Nature, UK.

Johannes, R. (ed.) 1989. *Traditional Ecological Knowledge: A Collection of Essays*. Gland, Switzerland, and Cambridge. International Union for the Conservation of Nature, UK.

Kimmerer, R.W. 2000. Native knowledge for native ecosystems. *Journal of Forestry* 98:4-9.

Mantua, N.J., and S.R. Hare. In press. The Pacific decadal oscillation. *Journal of Oceanography*.

Martinez, D. 1994. Traditional environmental knowledge connects land and culture. *Winds of Change* Autumn 89-94.

McCready, S. 2001. *The Discovery of Time*. Sourcebooks, Inc., Naperville, Ill.

Monson, D.H., D.F. Doak, B.E. Ballachey, A. Johnson, and J.L. Bodkin. 2000. Long-term impacts of the Exxon Valdez oil spill on sea otters, assessed through age-dependent mortality patterns. *Proceedings of the National Academy of Sciences of the United States of America* 97(12):6562-6567.

NRC (National Research Council). 1995. *Review of EPA's Environmental Monitoring and Assessment Program: Overall Evaluation*. National Academy Press, Washington, D.C.

NRC (National Research Council). 1996. *The Bering Sea Ecosystem*. National Academy Press, Washington, D.C.

NRC (National Research Council). 1999a. *Sustaining Marine Fisheries*. National Academy Press, Washington, D.C.

NRC (National Research Council). 1999b. *Global Ocean Science: Toward An Integrated Approach*. National Academy Press, Washington D.C.

NRC (National Research Council). 2000. *Ecological Indicators for the Nation*. National Academy Press, Washington, D.C.

NRC (National Research Council). 2001. *The Gulf Ecosystem Monitoring Program: First Steps Toward a Long-Term Research and Monitoring Plan*. Interim Report. National Academy Press, Washington, D.C.

Paine, R.T., J.L. Ruesink, A. Sun, E.L. Soulanille, M.J. Wonham, C.D.G. Harley, D.R. Brumbaugh, and D.L. Secord. 1996. Trouble on Oiled Waters: Lessons from the *Exxon Valdez* Oil Spill. *Annual Review of Ecological Systems* 27:197-235.

Pajak, P. 2000. Sustainability, ecosystem management, and indicators: Thinking globally and acting locally in the 21st century. *Fisheries* 25:16-25.

Park, P. 1993. What is participatory research: A theoretical and methodological perspective. Pp. 1-9 in P. Park, M. Brydon-Miller, B. Hall, and T. Jackson (eds.), *Voices of Change: Participatory Research in the United States and Canada*. Bergin & Garvey, London.

Park, P., and L.L. Williams. 1999. From the guest editors: A theoretical framework for participatory evaluation research. *Sociological Practice* 1:89-100.

Picou, J.S., and D.A. Gill. 1996. The *Exxon Valdez* oil spill and chronic psychological stress. Pp. 879-893 in S.D. Rice, R.B. Spies, D.A. Wolfe, and B.A. Wright (eds.), *Proceedings of the Exxon Valdez Oil Spill Symposium*, volume 18. American Fisheries Society, Bethesda, Md.

Rose, D. 1993. Reflections on ecologies for the twenty-first century. Pp. 115-118 in N. Williams, and G. Baines (eds.), *Traditional Ecological Knowledge: Wisdom for Sustainable Development*. Centre for Resource and Environmental Studies, Australian National University, Canberra.

Royer, T.C. 1993. High latitude oceanic temperature variability associated with the 18.6 year nodal tide. *Journal of Geophysical Research* 98:4639-4644.

Sustainable Biosphere Initiative. 1996. *Ecological Resource Monitoring: Change and Trend Detection*. Recommendations from a workshop held May 1-3, 1996, in Laurel, Maryland. Ecological Society of America, Washington, D.C.

Weisberg, S.B., T.L. Hayward, and M. Cole. 2000. Towards a US GOOS: A synthesis of lessons learned from previous coastal monitoring efforts. *Oceanography* 13(1):54-61.

APPENDIXES

A

Biosketches of the Committee's Members

Michael Roman, *chair*, is a professor at Horn Point Environmental Laboratories at the University System of Maryland's Center for Environmental Sciences. His research interests are biological oceanography, zooplankton ecology, food-web dynamics, estuarine and coastal interaction, and the carbon cycle in the ocean. Dr. Roman was chair of the Coastal Ocean Processes Steering Committee for the National Science Foundation and has experience leading a multidisciplinary activity. He brings a broad ecological perspective to this setting.

Don Bowen is a research scientist at the Marine Fish Division of the Bedford Institute of Oceanography's Department of Fisheries and Oceans in Canada. His research has focused on the population dynamics, foraging ecology, and ecological energetics of pinnipeds. Objectives of these studies are twofold: to understand the diversity of pinniped life histories and to understand the nature of competitive interactions between seals and commercial fisheries. Since 1997 Dr. Bowen has also conducted ecological research on the northern right whale with the aim to foster the recovery of the species.

Adria A. Elskus is an assistant professor of environmental physiology at the T.H. Morgan School of Biological Sciences at the University of Kentucky. Her scientific background includes work in endocrinology, geochemistry, biochemistry, and physiology, and she has worked as a consultant in industry, as a toxicologist and chemist in government, and in academia. Her research interests include the fate and effects of contaminants, including petroleum, in aquatic ecosystems, particularly effects on reproduction; adaptation to environmental contaminants; organic pollutant metabolism and the interplay of hormones and pollutants; and the biochemical mechanisms of pollutant effects. She also has specific experience in the analysis of samples collected from oil spill sites.

John J. Goering is a professor emeritus and former associate director of the Institute of Marine Science, University of Alaska, Fairbanks. He is well known as one of the first to make significant discoveries in the areas of the marine nitrogen cycle, the silicon cycle, and silicon and nitrogen assimilation by phytoplankton. He has served as vice-president and later president of the Pacific Section of the American Society of Limnology and Oceanography, as chair of the Oil Spill Recovery Institute Science Advisory Committee, and as a member of the North Slope Borough Science Advisory Committee and the Coastal Marine Institute Technical Advisory Committee.

George Hunt is a professor of ocean ecology at the University of California, Irvine. Dr. Hunt has published extensively on the foraging ecology of marine birds, mechanisms for trophic transfer to top predators in marine ecosystems and the impacts of oil spills on marine birds. He is currently investigating how climate variability can affect the control of energy flow in the Bering Sea. Dr. Hunt is a fellow of the American Association for the Advancement of Science and the American Ornithologists Union, and has previously served on the NRC's Committee on Mono Basin, (1985-1987), the Ecology Subcommittee of the Committee to Review Outer Continental Shelf Environmental Studies Program (1986-1992), and the Committee to Review Alaskan Outer Continental Shelf Environmental Information (1991-1994).

Seth Macinko is a assistant professor at the Department of Marine Affairs, University of Rhode Island. Previously he was a social and economic policy analyst at the Alaska Department of Fish and Game. He also fished commercially off Alaska from 1979 to 1983. His research interests are broadly focused on the interconnections between natural resource management (especially marine resources), environmental history, and political ecology. He is particularly interested in the role of institutional arrangements and culture in resource management. Current projects are focused on distributional issues involving access to marine resources property rights in marine fisheries, the role of place and community in property rights reformations, and linkages between marine resources and community development.

Donal T. Manahan is the director of marine biology at the University of Southern California. He is an environmental physiologist active in many areas of science in the Antarctic, as well as in temperate regions and deep-sea hydrothermal vents. His research includes physiological ecology of early stages (larvae) of animal development, animal/chemical interactions in the ocean, and the genetic bases of physiological processes. In education he is currently the director of an international Ph.D.-level training course in Antarctica, "Integrative Biology and Adaptation of Antarctic Marine Organisms." Dr. Manahan was the chair of the Polar Research Board from 1999 to 2002 and serves as the board's liaison to this activity.

Brenda Norcross is a professor of fisheries oceanography in the School of Fisheries and Ocean Sciences, University of Alaska, Fairbanks. Her research centers on fish and their habitats, including human-induced effects on the environment. She has studied flatfishes in Alaskan waters and has modeled nursery habitats. Dr. Norcross headed the herring component of the multi-investigator Sound Ecosystem Assessment project, which investigated the environment of Prince William Sound following the *Exxon Valdez* oil spill. That research resulted in a synthetic knowledge of the juvenile life stage of herring. She also has studied distribution of juvenile fishes and their availability to marine mammals, especially Steller sea lions.

J. Steven Picou is a professor of sociology and chair of the Department of Sociology and Anthropology, University of South Alabama. He is a leading authority on the social impacts of technological disasters and also has active research interests in clinical sociology and environmental sociology. From 1989 to 1992 he directed an interdisciplinary team of social scientists for assessing the community impacts of the *Exxon Valdez* oil spill. Dr. Picou also developed and implemented a clinical community intervention program in Cordova, Alaska, from 1994 to 1997 that was designed to reduce chronic, spill-related social and psychological impacts. At present, he is directing a long-term study of social consequences of the *Exxon Valdez* litigation and chronic ecological degradation in Prince William Sound, Alaska, and two projects on the health risks of consuming contaminated fish in the Mobile Bay Estuary in Alabama.

Tom Royer holds the Samuel and Fay Slover Distinguished Chair in Oceanography at Old Dominion University. Dr. Royer is a leading authority on the oceanography of the Gulf of Alaska. His research interests are in deep ocean and coastal hydrography and currents, longtime series measurements, and air-sea interactions. He was at the University of Alaska for several decades, where he was one of the cornerstones of their academic and research programs and where his discovery of a significant coastal current along the coast of Alaska, driven by freshwater discharge, allowed a reasonable prediction of the trajectory of the oil released during the 1989 *Exxon Valdez* oil spill. He represented the University of Alaska, Fairbanks in UNOLS for many years and led the UAF ship program. He has a very broad view of marine science, and he has seen extensive service on many panels, boards, and committees.

Jennifer Ruesink is an assistant professor of zoology at the University of Washington. Her areas of academic interest include community ecology, especially food-web interactions; species invasions; the conservation of biological diversity; and ecosystem functioning. She has studied the ecological impacts of the *Exxon Valdez* oil spill on the ecology of tidal communities in Prince William Sound, including work with National Academy of Sciences member Dr. Robert Paine.

Karl Turekian is the Silliman Professor of Geology and Geophysics at Yale University. He also is the director of the Yale Institute for Biospheric Studies and the director of the Center for the Study of Global Change. His research areas include marine geochemistry; atmospheric geochemistry of cosmogenic; radon daughter and man-made radionuclides; surficial and groundwater geochemistry of radionuclides; planetary degassing; geochronology based on uranium decay chain and radiocarbon of the Pleistocene; osmium isotope geochemistry; meteorite origins in relation to planetary systems; oceanic upwelling; and climate change. Dr. Turekian is an NAS member and has served on several NRC boards and committees including the Ocean Studies Board and the Committee on Global Change Research.

B

Acronyms

ACC	Alaska Coastal Current
ADCP	acoustic Doppler current profiler
BATS	Bermuda Atlantic Time Series
CalCOFI	California Cooperative Fisheries Investigations
EOS	NASA's Earth Observing System
EVOS	*Exxon Valdez* oil spill
EVOSTC	*Exxon Valdez* Oil Spill Trustee Council
GAK1	Gulf of Alaska station 1 located at the mouth of Resurrection Bay (60 N, 149 W)
GEM	Gulf Ecosystem Monitoring
GLOBEC	Global Ecosystem Dynamics program
GOA	Gulf of Alaska
GOFS	U.S. Global Ocean Flux Study
HOTS	Hawaii Ocean Time Series
IPRC	International Pacific Research Center
JGOFS	Joint Global Ocean Flux Study
LTER	Long-Term Ecological Research

NASA	National Aeronautics and Space Administration
NOAA	National Oceanic and Atmospheric Administration
NRC	National Research Council
NSF	National Science Foundation
ONR	Office of Naval Research
PSAMP	Puget Sound Ambient Monitoring program
PWS	Prince William Sound
RFP	Request for Proposals
RIDGE	Ridge Inter-Disciplinary Global Experiments
SALSA	Semi-Arid Land-Surface-Atmosphere program
SOLAS	Surface Ocean Lower Atmosphere Study